Smartphone Filmmaking

Smartphone Filmmaking

Theory and Practice

Max Schleser

BLOOMSBURY ACADEMIC
NEW YORK • LONDON • OXFORD • NEW DELHI • SYDNEY

BLOOMSBURY ACADEMIC
Bloomsbury Publishing Inc
1385 Broadway, New York, NY 10018, USA
50 Bedford Square, London, WC1B 3DP, UK
29 Earlsfort Terrace, Dublin 2, Ireland

BLOOMSBURY, BLOOMSBURY ACADEMIC and the Diana logo are trademarks of Bloomsbury Publishing Plc

First published in the United States of America 2021

Cover design: Eleanor Rose
Cover image courtesy of the author

Library of Congress Cataloging-in-Publication Data

Names: Schleser, Max, 1980- author.
Title: Smartphone filmmaking: theory and practice / Max Schleser.
Description: New York: Bloomsbury Academic, 2021. | Includes bibliographical references and index.
Identifiers: LCCN 2021002055 (print) | LCCN 2021002056 (ebook) | ISBN 9781501360336 (hardcover) | ISBN 9781501360329 (paperback) | ISBN 9781501360343 (ebook) | ISBN 9781501360350 (pdf)
Subjects: LCSH: Digital video. | Motion pictures–Production and direction. | Smartphones.
Classification: LCC TR860 .S3125 2021 (print) | LCC TR860 (ebook) | DDC 777–dc23
LC record available at https://lccn.loc.gov/2021002055
LC ebook record available at https://lccn.loc.gov/2021002056

ISBN: HB: 978-1-5013-6033-6
PB: 978-1-5013-6032-9
ePDF: 978-1-5013-6035-0
eBook: 978-1-5013-6034-3

Typeset by Deanta Global Publishing Services, Chennai, India
Printed and bound in India

To find out more about our authors and books visit www.bloomsbury.com and sign up for our newsletters.

Lucy, Elfie & Moritz

Contents

Acknowledgements

First, I would like to thank all of the filmmakers and creatives who are featured in this book and whose work was screened at the International Mobile Innovation Screenings. Through the Mobile Innovation Network and Association (MINA), I had the privilege to meet some most inspiring people and discuss mobile, smartphone and mobile filmmaking with them. While organizing a screening and festival means quite a bit of dedication and behind-the-scenes work, the collaboration with some fantastic colleagues made this a great experience. As MINA rebranded a few times, a special thanks to logo design by Thomas Le Bas. MINA is a mobile festival, which engages and collaborates with great institutions across New Zealand, Australia and China. In a chronological order, it was fantastic to work with Karen Curley, Klaus Kremer, Tim Turnindge in the College of Creative Arts at Massey University in 2011 and 2012; Laurent Antonczak at AUT University in 2013 and 2014; Associate Professor Dr Marsha Berry, Dr Smiljana Glisovic, Dr Seth Keen and Dr Patrick Kelly at RMIT University in 2015; Dean Keep at Swinburne University of Technology in 2016; Adrian Jeffs and Krish Nand at the Australian Centre for the Moving Image (ACMI) in 2017 and at Swinburne University of Technology in 2018; and Associate Professor Dr Xiaoge Xu at the University of Nottingham Ningbo China co-chairing the International Mobile Storytelling Congress, which featured the ninth International Mobile Innovation Screening and Festival. From 2011 to 2018 MINA screened at Ngā Taonga Sound & Vision, the New Zealand Archive of Film, Television and Sound, and it has been a pleasure to work with everyone at this institution, which was also my favourite cinema in Wellington, New Zealand / Aotearoa.

I would like to thank the interviewees in this book, especially Associate Professors Dr Camille Baker and Dr Gerda Cammaer. I had the pleasure to meet everyone in person at festivals, conferences, seminars or online. The continued support, discussions and their outstanding work are one of the inspirations for this book. I also would like to thank Swinburne University of Technology, especially the Department of Film and Animation, Department Chair Professor Dr James Verdon, Co-Chair Dr Antony Bartel, Course Director for Film & TV Dr Mark Freeman and Course Director for Screen Production Dean Keep, as well as the Centre for Transformative Media Technologies and Co-Directors Professor Dr Kim Vincs and Professor Dr Angela Ndalianis.

I also would like to acknowledge the Visible Evidence community where I presented my research into mobile-mentary (mobile documentary) filmmaking in 2015, 2012, 2010, 2008 and 2007 as well as the discussions on the Visible Evidence Listerv.

A special mention and big thank you to the reviewers and the team at Bloomsbury and especially Katie Gallof. I was first introduced to the Bloomsbury by my former PhD supervisor Professor Dr Joram Ten Brink at the University of Westminster in London, UK, in 2011. Also thanks to Thomas Insole who helped with reformatting the pictures and Miriam D'Oro for the amazing illustrations. Many thanks to everyone that contributed with insights, pictures and passion for smartphone filmmaking including numerous filmmakers, creatives, researchers, production companies and broadcasters.

And finally, I have to thank my family for their support throughout this book project.

This book is written by a smartphone filmmaker for those already part of this community and for those who seek to join the world of smartphone filmmaking. I hope it will provide some inspiration for future projects and create a record of what has already been produced, screened and celebrated by capturing more than a decade of mobile moving-image work. Every single smartphone film is worthy of appreciation as creative endeavour in shaping the global smartphone filmmaking scene. References include an extensive list of all smartphone films. All of these mobile and smartphone films are good as gold.

Illustrations

Figures

Table

1 Introduction

Mobile, smartphone and mobile filmmaking is a global phenomenon with distinctive festivals, filmmakers and creatives who are defining an original film form. *Smartphone Filmmaking: Theory and Practice* explores the diverse approaches towards smartphone filmmaking and provides an overview of the international smartphone filmmaking community.

This book not only develops a framework for the analysis of smartphone filmmaking but also reviews the contemporary scholarship and directions within the Creative Arts and the Creative Industries. *Smartphone Filmmaking: Theory and Practice* initiates a conversation on current trends and discusses its impact on adjacent disciplines and recent developments in emerging media and screen production, such as Mobile XR (extended reality). The interviews with smartphone filmmakers, entrepreneurs, creative technologists, storytellers, educators and smartphone film festival directors provide a source of inspiration and insight for professionals, emerging filmmakers and rookies who would like to join this creative community.

Once upon a time, stories were told exclusively in the movie cinema. Then came broadband and smartphones as well as new forms such as vertical video, new platforms such as TikTok, Instagram and Facebook Stories, which established new storytelling formats in the contemporary mediascape. While global 'streaming wars' shake up the film industry, independent filmmakers need to reinvent themselves to tell their stories in a twenty-first-century context. This research will exemplify the case of smartphone filmmaking as a creative intervention and mobilization of digital filmmaking. Many of the emerging technologies and platforms challenge conventional storytelling approaches and our understanding about narrative as it was defined in the last decades. We only see the tip of the iceberg with new micro and short-form content streaming services and disruptive technologies like blockchain and new story worlds in XR experiences. For the world of smartphone filmmaking, these developments are liberating and provide new opportunities. While a more conservative position might see these developments and new story forms and formats as villains, in *Smartphone Filmmaking: Theory and Practice* these are the protagonists driving creative innovation in the hero's journey of this research.

The Creative Arts research examined in this book emerges from within the smartphone filmmaking community. Max Schleser produced more than a dozen short and feature mobile and smartphone films, including *Max with a Keitai* and *Frankenstorm*. Since

2011 he curated the International Mobile Innovation Screening & Festival.[1] As mobile and smartphone filmmaking is a dynamic and fast-moving field of study, it is significant to provide a mobile, smartphone and mobile filmmaking–specific space for analysis. This book contributes to this endeavour that commenced with the world's first mobile feature film screening in the Lumiere Cinema on Regent Street in London (UK) in 2008.

As a starting point within an academic context one can observe the formations in the shaping of mobile, smartphone and mobile filming as a field in its own right through books like *Mobile Digital Art: Using the iPad and iPhone as Creative Tools* (Leibowitz 2013), *Smartphone Video Storytelling* (Montgomery 2018), *Smartphone Movie Maker* (Stoller 2017), *Hand Held Hollywood's Filmmaking with the iPad & iPhone* (Goldstein 2012), *Making Movies with Your iPhone: Shoot, Edit, and Share from Anywhere* (Harvell 2012) and mobile filmmaking guides such as *GoPro: Professional Guide to Filmmaking* (Schmidt and Thompson 2014). No publication to date has rigorously looked at mobile, smartphone or mobile filmmaking as a unique film form with its own qualities and characteristics. *Smartphone Filmmaking: Theory and Practice* explores smartphone filmmaking's aesthetic values and virtues, such as the significance of personal filmmaking ranging from political implications of representation to access to production tools and technologies, the notion of intimate filmmaking, the reference to the location and being close to the action and experience of the filmmaker. Chapter 6, 'Creative Innovation', examines the international renown smartphone feature films *Tangerine* and *Unsane*. While smartphone films are also produced in Hollywood with blockbuster budgets, screened at Festival de Cannes or Berlinale, this book will illustrate how filmmakers have broken rules of aspect ratio, image quality and narrative conventions to define creative innovation. The focus is on exploring new perspectives in form and content. Emerging formats and forms such as vertical videos, smartphone filmmaking specific aesthetics and authentic approaches to storytelling are discussed. This book also focuses on mobile moving-image arts and screen productions beyond the mainstream cinema and defines mobile and smartphone filmmaking with its specific qualities, dynamics and affordances.

Scholars like Camille Baker have pointed towards the mobile specificity in the context of performance in *New Directions in Mobile Media and Performance* (Baker 2018) or the edited collection *Intersecting Art and Technology in Practice: Techne/Technique/Technology* (Baker and Sicctio 2016). In a broader framework *Mobile Media Making in an Age of Smartphones* (Schleser and Berry 2018) and *Mobile Story Making in an Age of Smartphones* (Berry and Schleser 2014) explored the creative potential and the social innovation that storytellers, creatives and researchers demonstrated through engaging with an original mobile and/or smartphone-specific approach within the Creative Arts and adjacent disciplines. Other collections like *The Routledge Companion to Mobile Media* (Goggin and Hjorth 2014), *Creating with Mobile Media* (Berry 2017) or *Creative Mobile Media: A Complete Course* (Prasad 2017) analysed these developments around mobile media and smartphone creativity through media, communication and cultural studies or an ethnographic lens. In the documentary studies context, Helen De Michiel and Patricia R. Zimmermann's *Open Space New Media Documentary: A Toolkit for Theory and Practice* (Zimmermann and De Michiel 2017) provides a theoretical framework aligned with the

Figure 1.1 Mobile film festivals in 2012 (Schleser 2012).

aims and objectives of this book in analysing the collective new media practices of smartphone filmmaking (Figure 1.1):

> Open space configures networks not solely as digital interfaces, but as the nexus of people, places, and technologies. Open space documentary designs environment and interactions for ongoing dialogues that may or may not be resolved. Open space documentary practice reclaims technologies for people.
>
> (Zimmermann and De Michiel 2017, 144)

In 2019 the film festival submission platform FilmFreeway listed thirty-one festivals dedicated to smartphone filmmaking. Since mobile filmmaking surfaced more than a decade ago, it is time to discuss smartphone films through an original framework specific to the context of mobile, smartphone and mobile filmmaking. This book provides an overview of these developments and establishes five smartphone filmmaking specific modes. It is no longer possible to combine all approaches to smartphone filmmaking into one umbrella term without defining modalities as established in documentary studies (Nichols 2001) and more recently the field of i-docs, or interactive documentaries (Aston, Gaudenzi and Rose 2017).

Nichols's, as much as Aston, Gaudenzi and Rose's, framework identifies patterns in aesthetics that aligned with specific kinds of documentary and i-docs practice, respectively. The modes outlined here are not only a technical choice but also a creative approach. *Smartphone Filmmaking: Theory and Practice* studies innovation in form and content in filmmaking processes in production, distribution and exhibition.

Chapter 2, 'Towards a Theory and Practice of Mobile, Smartphone and Mobile Filmmaking', introduces five distinctive smartphone filmmaking modes and defines these as an original film form. Furthermore, the Creative Arts research method of curating with reference to the International Mobile Innovation Screening & Festival (2011–19) is outlined and positions smartphone filmmaking in the domain of experimental film and moving-image arts. This book is written from an insider perspective as a smartphone filmmaker and screening director of MINA, the Mobile Innovation Network and Association. This chapter also outlines the mobile and smartphone filmmaking projects and the collaborations that MINA established during the smartphone film festivals and screenings since 2011 (Figure 1.2).

Chapter 3, 'Spotlight on mobile and smartphone film festivals', provides an overview of the global smartphone filmmaking scene. Smartphone film festivals are a networking and knowledge transfer hub for the international smartphone filmmaking community. These festivals are key in celebrating and showcasing the constant innovation in smartphone filmmaking. As passionate as smartphone filmmakers are smartphone film festival organizers. Their work needs to be recognized beyond the applause at screenings and festivals. This chapter also contributes to the area of film festival research and engages Susy Botello, Si-Simon Horrocks, Angela Blake, Michael G. Osheku and Karl Bardosh in a conversation on their understandings and ideas regarding smartphone filmmaking. The International Mobile Film Festival is about accessibility and providing a space to present the achievement of producing smartphone films and presenting works to an audience and

Figure 1.2 Selected films for screening and festival (23–6 November 2011) at Ngā Taonga Sound & Vision, the New Zealand Archive of Film, Television and Sound, Wellington, New Zealand, Aotearoa (MINA 2011).

fellow smartphone filmmakers. Festival Director Susy Botello places particular emphasis on celebrating this with the red-carpet moment. MoMo – Mobile Motion – is positioned as a springboard and talent hub. Festival co-founder Si places the same focus on the community element, which is also crucial for SF3, SmartFone Flick Fest, which brings audiences and smartphone filmmakers together. In an interview with Angela Blake, she also talks about the introduction of smartphone feature films to the SF3 festival, continuing the accessible and affordable approach beyond the short film environment. ASIFF, African Smartphone International Film Festival, demonstrates the cultural dimension for localized content and African stories. Michael G. Osheku emphasizes the economic contribution with the introduction of a short film market. Karl Bardosh, as co-founder of the Indian Cell Phone Film Festival, discusses opportunities for the educational approach beyond screen and digital media as well as visual arts. For him, smartphone filmmaking is 'Hollywood in your Pocket', and he describes smartphone filmmaking as an equalizer. The impact of smartphone filmmaking goes beyond the film industry as a form of digital literacy in the twenty-first century (Figure 1.3).

Chapter 4, 'Focus on smartphone filmmaking practice', reviews the creative practices of renowned international smartphone filmmakers. These filmmakers and creatives developed original approaches towards smartphone filmmaking, and this chapter provides insights into their projects and productions. Gerda Cammaer's documentary art explores movement in all its meanings and forms. She describes her filmmaking approach as a bodily experience. Felipe Cardona's video-loops are based on repetition as a non-narrative strategy, which allows him to find a personal voice within these unique audiovisual experiences. Camille Baker's mobile media art explores haptics and wearables, investigating emotional and

Figure 1.3 Mobile, smartphone and pocket film festivals in 2018. Blue: smartphone festivals running for eight years, orange: five to seven years, green: two to four years and yellow: one year.

visceral qualities of mobile video. Through smartphone filmmaking, independent filmmaker Conrad Mess discovered his cinematic style, which is highly inspired and influenced by Quentin Tarantino. Mess is recognized as one of the most awarded iPhone filmmakers. While the approaches of filmmakers and artists are as distinctive as their backgrounds, it is key to point out that all creatives recognize the notion of mobility.

Moreover, this chapter, in sync with the previous chapter, demonstrates the contributions that Creative Arts research at universities internationally makes to this emerging Creative Arts sector. Next to their respective creative practices, Camille Baker and Gerda Cammaer are university educators who contribute to shaping and inspiring the next generations of smartphone filmmakers and creative professionals. Within this context one can also refer to Marsha Berry, Patrick Kelly and Dean Keep's Creative Arts research. Patrick Kelly's smartphone film *Give Me Faces and Streets!* is discussed in the context of the International Mobile Innovation Screenings (Figure 1.4).

His work illustrates how smartphone filmmaking can operate in online and offline spaces using smartphones not only as cameras but also as voice-recorders and recording devices of online environments such as Google Maps or social media platforms. Marsha Berry's mobile media artwork explores co-presence, wayfaring and virtual proximity using the camera phone as a compass (Kilby and Berry in Schleser and Berry 2018, 60) as exemplified in *Wayfarer's Trail* (Figure 1.5).

Dean Keep is creative practitioner who explores the ways in which visual media may inform our understanding of historical time, place and personal/cultural memories. His 2011 *Memory Cathedral* exemplifies how mobile media captures ephemeral sights, sounds and experiences of the everyday (Figure 1.6).

Figure 1.4 Still image: *Give Me Faces and Streets!* Filmmaker Patrick Kelly contributed to the International Mobile Innovation Screenings in 2013–18 and was also a member of the Screening Committee in 2014–18 and 2020 (Kelly 2016).

Figure 1.5 Still image: *Wayfarer's Trail* ethnographer and artist Marsha Berry contributed to the International Mobile Innovation Screenings in 2011, 2013 and 2016 and was also a member of the Screening Committee in 2013–16 (Berry 2016).

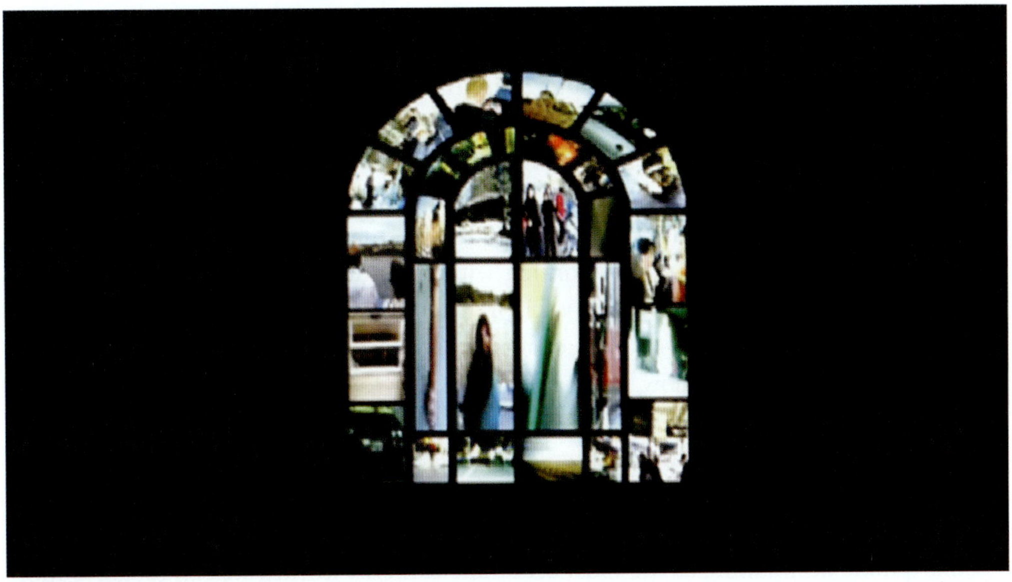

Figure 1.6 Still image: *Memory Cathedral.* Dean Keep contributed to the International Mobile Innovation Screening in 2011 and 2014 and was also a member of the Screening Committee in 2012, 2014–16 (Keep 2011).

Chapter 5, 'New horizons: Creative Industries', outlines developments in the Creative Industries. The smartphone filmmaking ecology includes production companies, software and app development, hardware R&D, government bodies and other cultural institutions. Terri Morgan from Luma Touch, Benoît Labourdette, a filmmaker, film festival organizer, educator and consultant to governments and cultural institutions, Adobe as a company with a mobile device strategy, Jason Van Genderen who runs Treehouse, an all-mobile device production company creating commercials and films on smartphones, and Aron Kumar from Struman Optics, reveal the various contributions smartphone filmmaking makes to the Creative Industries and beyond. Luma Fusion, the company that developed Luma Touch, attends IBC (International Broadcasting Convention)[2] in Amsterdam and the NAB show (National Association of Broadcasters) in Las Vegas,[3] and equally supports and contributes to mobile and smartphone film festivals and conferences. Benoît Labourdette emphasized that the Pocket Film Festival could secure appropriate budgets while maintaining an experimental approach and engage experimental filmmakers and mainstream audiences alike. Adobe Rush works with 4K video as device stability continues to extend to higher video resolutions. Adobe's engagement through events like VidCon,[4] 'the world's largest celebration of digital video & online video creators', demonstrates the formation of a specific online video ecology that smartphone filmmaking is contributing too. These developments are supported by filmmakers like Jason Van Genderen, through an entrepreneurial approach as 'lean-forward' filmmaking. Next to the contributions to the Creative Arts and the film industry, the interview with Struman Optics National Manager Aron Kumar points to the contribution smartphone filmmaking can make for the business environment and the broad application of video in trade and enterprise. While Chapter 5 outlines the entrepreneurial opportunities, Chapter 6 defines social innovation.

Chapter 6, 'Creative Innovation', outlines smartphone filmmaking's impact on independent filmmaking. This chapter discusses the new role of the 'director-cinematographer' as an extension from the 'writer-director'[5] and outlines innovation within independent filmmaking, including novel cinematic approaches in framing and composition, and emphasizes that mobile media art aesthetics still resonate in contemporary smartphone filmmaking. In addition, smartphone filmmaking facilitates new storytelling modes. It explores working with communities to co-create stories. The *Open Space New Media Documentary* theoretical framework (Zimmermann and De Michiel 2017) will be applied to the collaborative smartphone filmmaking approaches. This chapter also explores how experimentation drives innovation and points at the transformational capacity of creativity in collaborative smartphone filmmaking. In Chapter 6, Schleser outlines his Creative Arts research into smartphone filmmaking and discusses how his emphasis changed from an investigation into a visual to a collaborative aesthetic.

Smartphone filmmaking as the most accessible cameras worldwide can establish more democratic storytelling approaches with a drive towards equity. Smartphone filmmaking can allow for and facilitate bottom-up approaches and empower marginalized and underrepresented communities to provide an insider perspective about their situations and stories. As a growing ecology, including festivals (Chapter 3), practitioners (Chapter 4) and the Creative Industries (Chapter 5), smartphone filmmaking has an impact on the mediascape more generally and is significantly shaping the domain of online video. Within this context smartphone music videos are described as a new form of experimental film, giving (moving-image) artists and filmmakers the freedom for improvisation and experimentation. Driven by creativity, the independent and online sector established vertical video as a new video format with its own festivals. In order to illustrate how smartphone filmmaking modes, as defined in Section 2.6, find a wider resonance than the MINA screening and festival, this chapter reviews these modes and Smartphone filmmaking-specific formations, in the film, TV and digital media industries internationally. The final section, 'Mobile XR', will explore some of the most recent developments and speculate about the creative futures of smartphone filmmaking and demonstrates how it can mobilize the Creative Industries.

Chapters 3–5 feature interviews. These interviews demonstrate the multiplicity of approaches towards smartphone filmmaking. This book does not see smartphone filmmaking as a revolution but rather an evolution. While other books offer technical guides for mobile and smartphone filmmaking and apply conventional filmmaking approaches, *Smartphone Filmmaking: Theory and Practice* curates the space within the domain of moving-image arts and discusses storytelling and aesthetic approaches beyond the mainstream. The affordances of smartphone filmmaking can further expand traditional industry conventions. The notions of mobility and accessibility define an original capacity specific to mobile, pocket and smartphone filmmaking.

This book particularly highlights smartphone filmmaking in the context of experimental filmmaking and screen production as an innovative film form. It further develops a cinematic language, filmic grammar and our understanding of what a story is and will be in the near future. If you like it or not, one of the key characteristics of digital media, screen production and emerging media, is its continuous innovation, reformation and rapid morphing into new forms and formats. The handbooks mentioned earlier merely translate filmmaking 1:1, meaning 'one-to-one', without understanding the different syntax, technical and sociocultural dimensions that are interacting with smartphone filmmaking. The last 2010s showed us a distinctive visual aesthetic that paved the way for a broader understanding of mobile and smartphone filmmaking's possibilities. While not every story is appropriate to be realized with a pocket camera or smartphone, if working with communities, capturing locations or working in the domain of personal or first-person filmmaking, the smartphone, pocket or mobile device should be considered as the camera of choice. The mobile specificity is here expressed through its intimate and immediate qualities, which are further discussed in this book. These characteristics, personal forms of crafting experiences and sharing these online, contribute to a formation of new storytelling forms, moving from a script to screen to shoot and share approach. Furthermore, smartphone filmmaking's characteristics have an impact on filmmaking, industry conventions and film language more generally. Stylistic developments of vertical video and collaborative processes in smartphone filmmaking are evolving into hybrid forms and formats that resonate in other film forms.

As this book defines smartphone filmmaking over a decade, the terminology 'mobile, smartphone and mobile filmmaking' is chosen. One approach could differentiate mobile and smartphone filmmaking based on its technical development in video codecs and standards. The pioneers (see Section 2.1) used 3GP video files which operated with QuickTime video dimensions at about 128 × 96, 176 × 144 or 352 × 288 pixels. With the introduction of the iPhone 4 (2010) 720p and iPhone 6 (2014) 1080p new video formats were established. In the first MINA screening in 2011, Nokia was the choice of camera phone by most mobile filmmakers, while now most smartphone filmmakers use Apple's iPhone. Next to the improvement in sensors and video codecs, it is key to consider the entire 'picture'. Apple's App Store was launched in 2008 with 500 applications (Bonnington 2013) and by 2018 a new sector in digital enterprise emerged, generating $86 billion in sales for iOS developers over ten years (Apple 2018). Smartphone filmmaking is positioned at the intersection of the new digital economy, creativity and community engagement. Chapter 5 evidences the opportunities for app developers, gear manufacturers, content creators, cultural consultants and digital storytellers to be part of this new ecology. In this context Treehouse is one of the first 'all-mobile' production companies, whose work spans across several formats and commercial productions. Here, the creativity and craftsmanship demonstrate that smartphone filmmakers can do the work of standard broadcasting cameras, while it is not the other way around.

Next to the economic dimension, smartphones facilitate production approaches without desktop computers. The 'Smartphone Native Mode' (Sections 2.6.3. and 6.4.4.), as much as the interview by Felipe Cardona (Section 4.3), illustrates this smartphone-specific aesthetic development. Key in these approaches is mobility. Smartphone filmmaking further developed the mobile filmmaking's specificity. The notion of mobility is discussed in the interviews with Cammaer (Section 4.2), Baker (Section 4.4) and Morgan (Section 5.2), as well as in the context of music videos (Section 6.3.2) and independent filmmaking (Section 6.4.1) exemplified through the feature productions *Blue Moon* or *Char Man*. When considering the possible futures of smartphone filmmaking in the context of *Mobile XR* (Section 6.7), mobility remains key. The other significant consideration is the accessibility of smartphone filmmaking; at this point in time smartphone filmmaking is the most egalitarian filmmaking form. As NYU educator and filmmaker Karl Bardosh points out, smartphone filmmaking is an equalizer (Section 3.6). Mobile, smartphone and mobile filmmaking can create equity in production environments and with the potential of social media for distribution and exhibition, this digital ecology provides new opportunities that operate in novel ways in comparison with more traditional filmmaking modes. Creatively, smartphone filmmaking provides prospects to expand the boundaries in the field and demonstrate alternative practices as outlined in Sections 2.6 and 6.4.

Furthermore, it is significant to emphasize that the aesthetic qualities, which *the Pioneers* established in mobile filmmaking, which was refined by the second wave of mobile filmmakers (2008–12), still resonate in contemporary smartphone films. This 'Keitai aesthetic' (Schleser 2011, 93) is linked to mobility in mobile, smartphone and mobile filmmaking. Chapter 2 defines the meaning of Keitai and points to the mobility in its definition: 'small, portable, carrying, something, form, shape or figure' (Schleser 2011, 93). Smartphone filmmaking is embracing these early mobile filmmaking qualities and mobility continuous to define contemporary smartphone filmmaking practices. As emphasized in this book, not for every film would be a sound smartphone film, but the smartphone is the camera of choice when telling personal stories, filming on a journey on location or in life, working with communities or aesthetic explorations. The term 'mobile filmmaking' is adapted from the French *Pocket Film Festival*, which was held in the Forum des Images in Paris from 2005 to 2010.[6] The French mobile film festival was organized by Benoît Labourdette, who is interviewed in Chapter 5.

This book positions smartphone filmmaking as a continuation of alternative practices that emerged out of a filmmaking community. Historical precedents include Kino Pravda (Vertov in Michelson 1984) and the French New Wave (in the late 1950s and early 1960s) among other filmmaking forms that explored new ideas in rejection of traditional approaches or industry conventions. Smartphone filmmaking provides new opportunities to expand on these filmmaking traditions. At the time of writing this book, more and more smartphone feature films are being produced and screened at smartphone film festivals (Chapter 3) internationally. Producing a smartphone film means that budget constraints are less significant and if there is a budget this can be spent in different ways (i.e. enabling to travel to a specific location rather than to spend 'dollars' on camera rental or purchase). As there is now a recognition of smartphone filmmaking at major international film festivals as outlined in Chapter 6, it is even more prominent to point out that these films used smartphones as cameras which could not have been replaced by standard broadcasting cameras due to the smartphone's mobility and aesthetic choices. Next to the video codec standards, audio is a key consideration for major international film festivals. There are now several microphones on the market that facilitate the needs and budget requirements of every smartphone filmmaker. Even if one does not have a microphone, a second smartphone can be used as an audio recorder in order to be as close to the sound source as possible. Also, most smartphone headphones have a microphone these days. Of course, any microphone is better than no microphone, but this could be one of the first accessories to add to a smartphone filmmaking kit. There is no right or wrong answer to the question if external lenses should be used in the world of smartphone filmmaking. Still, this book makes the point about mobile filmmaking, if one attaches a lens to a smartphone that costs more than the smartphone and uses DOF (depth-of-field) adapter with a SLR, DSLR or cine lens, the specific smartphone filmmaking experience will be changed conceptually in terms of mobility as well as cinematography.

Moreover, the mobility of smartphone filmmaking with a big lens would be significantly limited. Certain tools, such as Gimbals, further develop a particular aesthetic. While for some film projects a well-practised and rehearsed handheld camera movement provides a more energetic element, in other situations, a more spontaneous camera move can create authentic and empathic engagement. Depending on the smartphone filmmaking project, Gimbals can be a fantastic stabilization tool, that is recording interviews in a car or dynamic camera movements in music videos.

For *Smartphone Filmmaking: Theory & Practice* any 'accessory' like microphones, lenses, neutral density (ND) filters and so forth that would fit into one's pocket and thus are not substantially larger than the smartphone, should be included in the technical and aesthetic toolkit of smartphone filmmaking. In addition, there are a number of rigs such as Shoulderpod or Blackwing that allow for additional assessors to be mounted next to the smartphone camera, such as microphones and/or lights to small-scale smartphone filmmaking set-ups with various lenses. As long as these fit into a pocket, I would argue, they are a useful addition to mobile and smartphone filmmaking. There are several options from DIY approaches to high-end gear manufacturers who now produce dedicated mobile lighting, lenses and ND filter equipment. Smartphone filmmaking's mobility speaks to productions on location, and one could argue that the best results are achieved with natural light on location rather than in the studio anyhow. ND filters are like sunglasses for your smartphone and allow filmmakers to improve exposure in bright environments. Circular polarizing (CPL) filters reduce reflections from surfaces such as water or glass and provide more contrast in the image. For a cinematic look, the Moondog Labs anamorphic lens is used by many iPhone filmmakers whose projects are featured in this book. Another example of mobile filmmaking accessories is the small lens kit by Struman Optics, including a wide-angle, macro and fisheye lens. Beyond the world of storytelling, a macro and wide-angle lens can provide useful additions for trades and businesspeople documenting and showcasing their work. Video is now a twenty-first-century literacy and is becoming more and more an important communication format. Again, every project has its specifics, preferred camera styles and works with certain conventions or aims at breaking these. The magic of smartphone filmmaking is that one can create and experiment. Any standards or gear should not limit this. Susy Botello sums this up in her interview saying that 'the reason you and I and others created these film festivals was to give an opportunity to everybody that had a phone to realize their dream and get their foot in the door through smartphone filmmaking'. We would not want to see these 'add-ons' as stopping anyone from approaching this creative community. More significant than the choice of gear is the choice of collaborators, locations or aesthetic explorations. Every project has its own specificity and can operate within certain parameters. Thus, not having a DSLR camera, ARRI, Red or Blackmagic camera should not stop any producer or film commissioner in questing a production. Depending on the project, budget and experience of filmmakers, these 'filmmaking accessories' can be useful additions for a smartphone filmmaking kit. Most smartphone filmmakers featured in this book use the cinematic video app FiLMiC Pro. The app also facilitates more advanced cinematographic options, such as manual shutter speed, ISO, focus or different frame rates, to work flat or log v2 mode (short for logarithmic) which provides more dynamic range and thus greater tonality. In combination with dedicated camera profiles such as Filmconvert, it offers even more options in the post-production to meet the various aesthetic demands and cinematic looks. And in combination with apps like Luma Fusion or Adobe Rush, one can not only film but also edit with a smartphone, mobile device or tablet. Sections 2.6.3 and 6.4.4 discuss the opportunities for the production of short content on the go.

A new way of filming also facilities new editing approaches. Chapter 6 demonstrates how this continuum merges to the smartphone native mode. The study of film production in its sequencing of production, distribution and exhibition is now fundamentally altered. New production modes are facilitated by digital distribution forms and formats that now gain more and more significance. Within this digital ecology, the shift from script to screen to 'sh%t & share' creates new experiences in producing as much as watching smartphone films in various modes. Next to being the most used camera for online pictures and videos, smartphones now also contribute to the formation of new story forms and formats. Next to the video platforms YouTube and Vimeo, micro-content on video-sharing platforms

Facebook and Instagram Stories, Twitter or TikTok, one can observe the emergency of vertical video streaming services which dramatically alter traditional viewing formats. Examples include Quibi in the United States, which has signed Hollywood director Steven Soderbergh and secured investments from Hollywood studios (Spangler 2019); Vertical Motion Studio (by the German Broadcaster ZDF); and vertical feature film productions (i.e. the Russian feature film *V2. Escape From Hell*). Next to the vertical aspect ratio, which defines its departure from traditional cinematic conventions into new domains, one can observe the formation of a new profession of screen producers, the online and digital video creators at congresses such as VidCon (including the 'Stars of Tik Tok'), new production hubs, for instance, YouTube Spaces and new apps/software such as Adobe's new editing software Adobe Rush, created for online video and smartphone filmmaking to edit on the go. In the domain of social and networked media, smartphone video challenges cinematic narrative conventions. These novel video formats further develop transmedia storytelling approaches. Smartphone filmmaking now morphs across various screens and platforms.

Smartphone filmmaking is driven by the creative experimentation of filmmakers, storytellers as well as creative technologists and can thus be seen as a continuation of filmmakers that worked with Super 8 or digital filmmaking. As much as Super 8 in the 1980s and digital filmmaking at the turn of the century, early mobile filmmaking developed in an arts scene and community outside an industry context.[7] The creative experimentation and aesthetic exploration can also be linked to Joram Ten Brinks's *The Man Who Couldn't Feel* (1996). As the Super 8 feature film was produced in the academy, creative practice research demonstrates how a theoretical framework can emerge out of the work itself and in this case defined the essay film as an independent genre. Referring back to Super 8, Mike Figgis says in his book *Digital Filmmaking* (2007):

> from an aesthetic point of view, with Super-8 you immediately got a result that was representational but also interestingly removed from reality – you were looking at something that had a kind of richness and artistic potential to it.
>
> (2007, 30)

This modality can be translated to the Keitai aesthetic and smartphone filmmaking. Mike Figgis is quoted here, as he also produced the mobile film *Life Captured* (2008), which depicts 'everyday happenings and personal moments, providing a window into people's private worlds previously unseen' (Figgis 2008 online). He is more known for his digital filmmaking, such as the experimental film *Timecode* (Figgis 2000). As the film credits reveal, *Timecode* was filmed in four continuous takes for 93 minutes with four digital video cameras, maxing out the digital video cameras' storage capacity at 93 minutes. Apple commissioned Axinya Gog in a similar project with *Hermitage*, which was filmed on an iPhone 11 Pro 'in one continuous take on one battery charge'[8] (Apple 2020), a la *Russian Ark* (Sokurov 2002) in the world's largest museum, the Hermitage in St. Petersburg, Russia.

Smartphone filmmaking is operating in this digital filmmaking mode and allows to expand on filmmaking in form, format and content. With accelerated technical development in smartphone filmmaking, it is key to understand how an original film form was shaped by the material and creative practice by independent filmmakers. In order to emphasize the community element in smartphone filmmaking, this book included the mobile and smartphone film festivals within its analysis. Chapter 3 reviews the smartphone filmmaking scene on every continent (with the exception of Antarctica, where Michaela Skovranova created *Uncover Antarctica*, through the lens of an OPPO Find X2 Pro smartphone for National Geographic). Smartphone filmmaking operates with a fixed F-stop, the iPhone 7 at 1.8 and the iPhone 11 at 1.8 or 2.4 depending on the choice of camera, which means that as a filmmaker one always has to be as close as possible to the action, scene or environment being captured. This limitation is simultaneously an affordance (Norman 1998) creating an original experience; one is so close to the action, scene or environment that one could touch it.

Smartphone Filmmaking: Theory and Practice resulted out of practice-led research as a filmmaker and curator. Through the recognition of novel processes, production approaches and the continued aesthetic refinement in smartphone filmmaking, an original framework for analysis is established.

Notes

1. International Mobile Innovation Screening, https://www.mina.pro/
2. International Broadcasting Convention, https://www.ibc.org/
3. National Association of Broadcasters, https://nabshow.com/2020/
4. VidCon, https://www.vidcon.com/
5. Such as 'Writer-Directors' featured on IMDb (Internet Movie Database), https://www.imdb.com/list/ls068983855/
6. Pocket Film Festival, https://www.benoitlabourdette.com/actions-culturelles-et-pedagogiques/conception-d-evenements-culturels/festival-pocket-films/catalogues-de-toutes-les-editions-du-festival-pocket-films?lang=en
7. The Berlin Underground filmmaking scene in the late 1970s and early 1980s represents a snapshot in time that can be compared to the contemporary smartphone filmmaking scene. In 1981 network artist, activist and filmmaker padeluun brought together the independent film and arts scene under the theme 'Alle Macht Super 8', which translates to all power or force to Super 8. The documentary *Alle Macht der Super 8* captures this scene between 1978 and 1981. The rhyme was also a political slogan and was inspired by the Berlin Punk counterculture, which included international artists like David Bowie, Nick Cave or Iggy Pop.
8. Apple's #ShotoniPhone *Hermitage: 5 hrs 19 min 28 sec in one continuous take – official trailer*, https://youtu.be/aLLsmJfvixM

References

Apple. 2018. *App Store Kicks Off 2018 with Record-Breaking Holiday Season*. Available online: https://www.apple.com/newsroom/2018/01/app-store-kicks-off-2018-with-record-breaking-holiday-season/.

Aston, Judith, Gaudenzi, Sandra and Rose, Mandy, eds. 2017. *I-Docs: The Evolving Practices of Interactive Documentary*. New York: Wallflower Press.

Baker, Camille. 2018. *New Directions in Mobile Media and Performance*. New York: Routledge.

Baker, Camille and Sicctio, Kate, eds. 2016. *Intersecting Art and Technology in Practice: Techne/Technique/Technology*. New York: Routledge.

Berry, Marsha. 2017. *Creating with Mobile Media*. London: Palgrave Macmillan.

Berry, Marsha and Schleser, Max, eds. 2014. *Mobile Story Making in an Age of Smartphones*. London: Palgrave Macmillan.

Bonnington, Christina. 2013. '5 Years On, the App Store Has Forever Changed the Face of Software', *Wired Magazine*. Available online: https://www.wired.com/2013/07/five-years-of-the-app-store/.

Figgis, Mike. 2007. *Digital Filmmaking*. New York: Farrar, Straus and Giroux.

Goggin, Gerhard and Hjorth, Larissa, eds. 2014. *The Routledge Companion to Mobile Media*. New York: Routledge.

Goldstein, Taz. 2012. *Hand Held Hollywood's Filmmaking with the iPad & iPhone*. San Francisco: Peachpit Press.

Harvell, Ben. 2012. *Making Movies with Your iPhone: Shoot, Edit, and Share from Anywhere*. Hachette: Ilex.

Leibowitz, David. 2013. *Mobile Digital Art: Using the iPad and iPhone as Creative Tools*. New York: Routledge.

Montgomery, Robb. 2018. *Smartphone Video Storytelling*. New York: Routledge.

Nichols, Bill. 2001. *Introduction to Documentary*. Bloomington: Indiana University Press.

Norman, Donald. 1998. *The Design of Everyday Things*. New York: Basic Books.

Prasad, Sylvia, ed. 2017. *Creative Mobile Media: A Complete Course*. London: World Scientific Publishing.

Schleser, Max. 2011. *Mobile-Mentary: Mobile Documentaries in the Mediascape*. Saarbrücken: LAP Lambert Academic Publishing.

Schleser, Max and Berry, Marsha, eds. 2018. *Mobile Media Making in an Age of Smartphones*. London: Palgrave Macmillan.

Schmidt, Bradford and Thompson, Brandon. 2014. *GoPro: Professional Guide to Filmmaking*. San Francisco: Peachpit Press.

Spangler, Todd. 2019. 'Tye Sheridan Starring in Survival Thriller 'Wireless' from Steven Soderbergh for Quibi', *Variety*. Available online: https://variety.com/2019/digital/news/quibi-wireless-tye-sheridan-steven-soderbergh-1203405209/.

Stoller, Michael. 2017. *Smartphone Movie Maker*. Somerville: Candlewick Press.

Vertov, Dziga in Michelson, Annette, eds. 1984. *Kino-Eye*. The Writings of Dziga Vertov. Berkley: University of California Press.

Zimmermann, Patricia and De Michiel, Helen, eds. 2017. *Open Space New Media Documentary: A Toolkit for Theory and Practice*. New York: Routledge.

Films

Alle Macht der Super 8. 2005. Dir. Berlin House of Media. Germany: Alive.

Blue Moon. 2018 . Dir. Stefen Harris. New Zealand: Dark Horse Films.

Char Man. 2019. Dir. Kurt Ela and Kipp Tribble. United States: MRP Entertainment, GK & K Productions.

Frankenstorm. 2012. Dir. Max Schleser. United States: independent production.

Give Me Faces and Streets!. 2016. Dir. Patrick Kelly. Australia: independent production.

Hermitage: A One-Take Journey through Russia's Iconic Hermitage Museum. 2020. Dir. Axinya Gog. Russia/United States: Apple.

Life Captured. 2008. Dir. Mike Figgis. United Kingdom: Sony Ericsson.

Max with a Keitai. 2006. Dir. Max Schleser. Japan: independent production.

Memory Cathedral. 2011. Dir. Dean Keep. Australia: independent production.

Russian Ark. 2002. Dir. Sokurov. Russia: Egoli Tossell Film AG and Hermitage Bridge Studio.

Tangerine. 2015. Dir. Sean Baker. United States: Magnolia Pictures.

The Man Who Couldn't Feel. 1996. Dir. Ten Brink. United Kingdom: J. Productions.

Timecode. 2000. Mike Figgis. United States: Screen Gems and Sony Pictures Releasing.

Uncover Antarctica. 2020. Dir. Michaela Skovranova. United States: National Geographic.

Unsane. 2018. Dir. Steven Soderbergh. United States: Regency Enterprises and Extension 765. 20th Century Fox.

V2. Escape From Hell. In production. Dir. Timur Bekmambetov. Russia: Bazelevs, Voenfilm and MYS Media.

Wayfarer's Trail. 2016. Dir. Marsha Berry. Australia: independent production.

2

Towards a theory and practice of mobile, smartphone and mobile filmmaking

Chapter Outline

When considering mobile, smartphone and mobile filmmaking as a film form in its own right, one can compare and contrast smartphone films among each other. This analysis is common when a particular movie, genre or time period is discussed among other films in the same genre. Through establishing a theoretical framework specific to mobile, smartphone and mobile filmmaking, a conversation regarding the understanding of smartphone filmmaking specificity can be formulated. In the same way that an impressionistic painting, a visual communication design project, a constructivist photograph or a first-person documentary film is studied in the context of other artworks, photos, designs or movies within the same art, design or film form, movement or style, mobile, pocket and smartphone filmmaking is explored as an original film form. While the links to and inspirations from other periods and filmmaking styles suggest themselves as presidents and demonstrate a continuation of the respective style, mobile, smartphone and mobile filmmaking has the potential to provide innovative approaches and, as outlined in Chapter 5, disrupt the Creative Industries processes as well as markets. Furthermore, a number of large-scale format camera manufacturers are realizing the growth and potential of mobile filmmaking and in response launching small mobile cameras, such as Blackmagic Design[1] or ARRI,[2] and in the case of RED[3] even producing a smartphone. The first narrative short film *N.I.* by Sally Massimini was shot on a RED Hydrogen prototype phone in 2019[4].

Figure 2.1 Festival programme: Pocket Film Festival 2005 (Labourdette 2005).

The innovation from a technical viewpoint is driven by these companies, from an artistic position, and these developments are defined by the filmmakers outlined in this book. It is critical to understand that the discussion around mobile media as much as smartphone filmmaking can operate outside a traditional analysis. Over the last decade, there was a proliferation of short films, documentaries, mobile-mentaries (mobile documentary)

(Schleser 2011), Cellphilms (MacEntee, Burkholder and Schwab-Cartas 2016), mobile cinema (Atkinson 2017) and mobile filmmaking (Berry 2017). The contextualization is driven by the scholar's backgrounds and demonstrates the impact of smartphone filmmaking on adjacent disciplines. Cellphilms is a participatory visual method based on community-based research and activism (MacEntee, Burkholder and Schwab-Cartas 2016). Berry explores co-presence, the intermediate between online and offline. Her analysis focuses on the cultural meaning of sharing moments and outlines the phenomenological dimension through a lens of digital ethnography. *Smartphone Filmmaking: Theory and Practice* further expands into the creative process in filmmaking, an investigation into mobile, pocket and smartphone filmmaking in the context of storytelling and its engagement with distinctive visual and aesthetic parameters in film, TV, screen production and digital media. Sarah Atkinson's *Mobile Cinema* introduces an interesting distinction through films made on mobile phones, movies made for a mobile viewing experience and interactive films, which use mobile devices to guide audiences through story worlds via apps. She discusses films like *Night Fishing*, *Olive* and *Sotchi 255*, which are also mentioned in this book but does not engage with the independent or moving-image arts domain. Not mentioning these works does not mean that these do not exist. Especially in the context of mobile, pocket and smartphone filmmaking, this is key as mobile filmmaking emerged out of this domain. The idea of films made for mobile devices such as *Rage*, *The Sliver Goat* or *Haunting Melissa* is not discussed in this book as they are not produced entirely on smartphones or mobile devices. Another distinction this book makes in the transmedia projects discussed in Chapter 6 is the defining element of the collaborative process, which includes smartphone videos from audiences and thus moves beyond the entertainment (and media consumption) factor as exemplified by Atkinson with *App* or *RIDES*. Within this dynamic field, the most common characteristics within mobile, smartphone and mobile filmmaking are explored with a particular focus on production as well as creative processes.

2.1 The pioneers

Although mobile filmmaking has been around for more than a decade, smartphone filmmakers are continuously defining a new film grammar and continue to expand the boundaries in filmmaking, moving-image arts and the dynamic environment of screen production. Through studying the pioneers of mobile filmmaking and relating the characteristic of their productions to more contemporary smartphone films, the presented analysis forms a conceptual schema, which is outlining an emerging filmmaking form with a still very young history. The first mobile camera phone was introduced in Japan in 2000 (Schleser 2014, 156) and from 2004 onwards the first exhibitions and screenings featured micro-movies and mobile phone films (Figure 1.1). In 2008 the FILMOBILE exhibition (Schleser 2008) in the London Gallery West and the screening in the Regent Street Lumiere Cinema (London) featured the world's premiere of a dedicated mobile feature film screening including *SMS Sugar Man*, *Nausea* and *Max with a Keitai*. The event is featured in the experimental documentary *By Any Old Light*. In the film Aryan Kaganof, director of *SMS Sugar Man* and 1960s experimental filmmaker Peter Whitehead discuss Whitehead's legacy in experimental film and share their views on mobile, pocket and smartphone filmmaking. Aryan Kaganof continued to work with mobile and smartphone filmmaking and in 2019 he co-produced the project *Herri*, which explores 'what does decolonisation looks like in this age of hybridity' (Kaganof online 2019). *MBE MBHELE Uma umama engeneme ingane iyezwa egzzini – When a Mother Cries Her Children Feel It* is a smartphone documentary presented in the form of African storytelling as a song in isiZulu language. He is currently working on a new project with the Huawei P30 Lite smartphone.

The city film *Max with a Keitai* (Figure 2.2) applied interval editing in order to create a specific form to synthesize about 180 hours of filmed mobile phone videos and linked these from a web blog into a city film. The experimental documentary was shot entirely on two mobile phones in Japan in 2006. *Max with a Keitai* was screened in the Museum of Moving Image in San Paulo, the Pocket Film Festival (see also Figure 2.7) and is part of the public film archive in the Forum des Images in Paris. *Max with a Keitai* explores Japanese metropolitan centres through the lens of a mobile phone and captures an emerging mobile phone video aesthetic, which surfaced in and characterizes the years 2005–8. The city film captures the everyday life of the mobile phone filmmaker Max during the mobile-mentary production and the Japanese megapolis in the Taiheiyō Belt (Figure 2.3). The cityscapes are depicted as a hybrid of tradition and progressive technoculture. In addition, *Max with a Keitai* provides an alternative reading of the technologically most advanced centre. Max recorded the failures of

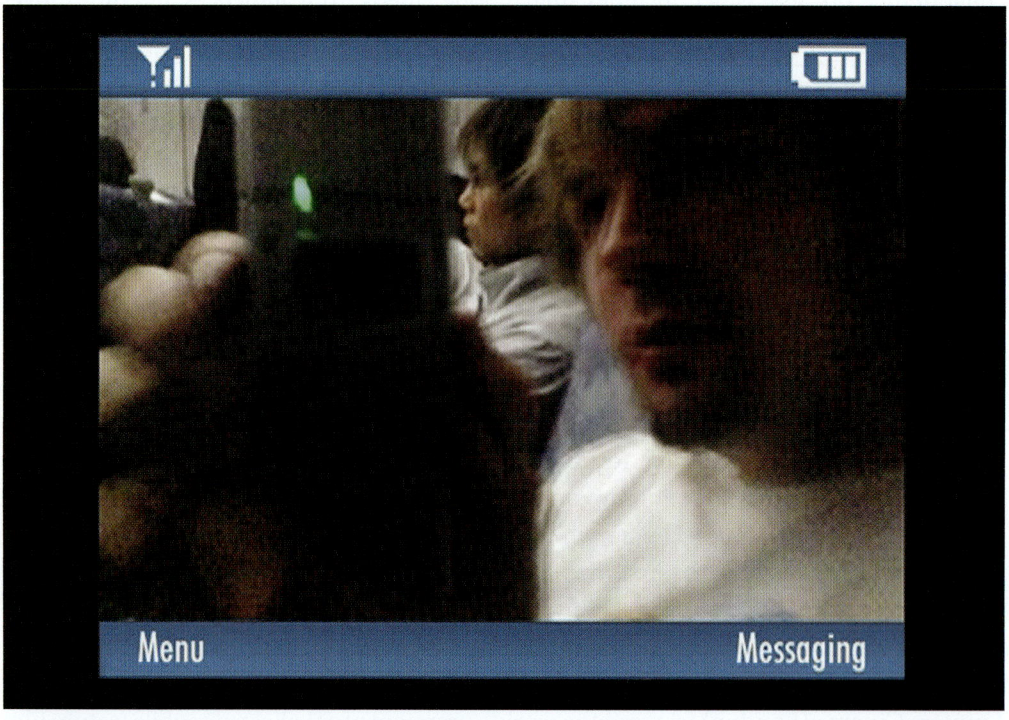

Figure 2.2 Still image: *Max with a Keitai* (Schleser 2008).

the technoculture, such as the derelict shopping mall in Den-Den Town (Electric City) with his keitai (mobile phone). A short of the experimental city film *Max with a Keitai* was edited on location and screened for the first time in Japan at the Design Fiesta in Tokyo in December 2006. The screening of the film to a Japanese audience became part of the feature project and is a direct reference to Kinoks (Vertov 1922) filmmaking.

This book positions mobile, smartphone and mobile filmmaking as an opportunity to update experimental film and avant-garde approaches in the twenty-first-century context. In a similar way that short films have long been a testing ground and innovation sandbox for filmmaking more generally, one can now observe that the Creative Industries are recognizing and embracing the online and vertical video world. As outlined in Section 6.3.1, while being ignored for a long time, one can now find the vertical video film festivals, and vertical video is about to be recognized as a new video format and aspect ratio. Vertical video gained popularity by advertisers, streamers and audiences alike. To understand the mobile-specific characteristic, such as intimate qualities and the significance of personal filmmaking, it is key to study the mobile filmmaking pioneers.

SMS Sugar Man (Figure 2.4) is a narrative feature film shot on the Sony Ericsson W900i in 2006/2007. Aryan Kaganof is director and protagonist in the film itself. As the name suggests, the story reveals what happens at Christmas Eve in Johannesburg's luxury hotels. The mobile phone also drives the narrative through text messages and opens up the

Figure 2.3 Still image: *Max with a Keitai* (Schleser 2008).

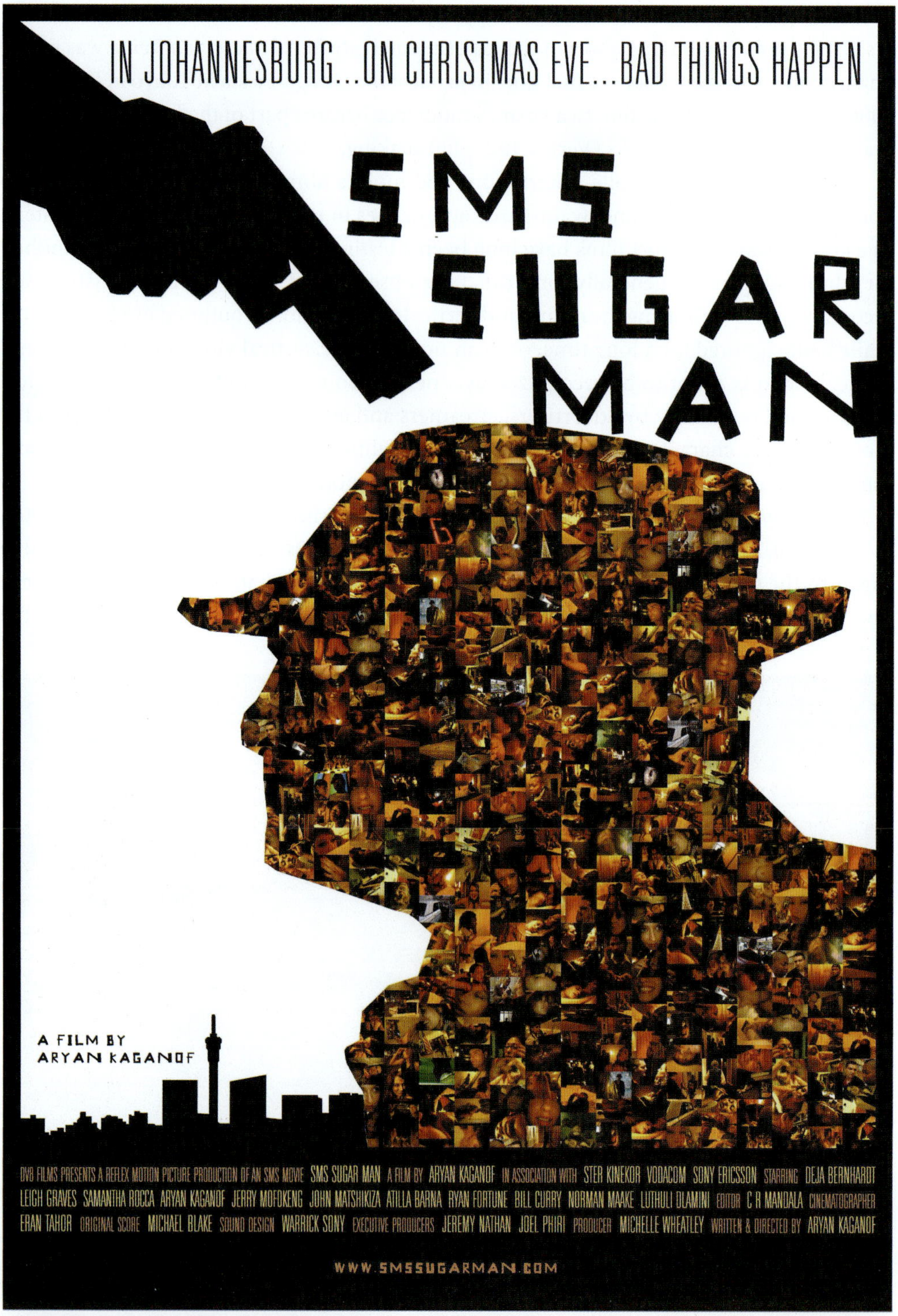

Figure 2.4 Film poster: *SMS Sugar Man* (Kaganof 2007).

intimate space of pimps and prostitutes. The protagonist's interactions and emotions are revealed through extreme close-ups and personal commentary. 'From a technical perspective, one should add that the accomplishment of SMS Sugar Man depends upon the way in which the film exploits the strengths of the mobile video whilst overshadowing the unique look and feel of the pixel aesthetic' (Schleser 2011, 44) (Figure 2.5).

This pixel aesthetic was not only a visual defining characteristic but also mirrored the personal storytelling approaches as seen in *Max with a Keitai* or *Nausea*. The experimental mobile film by British filmmaker Matthew Noel-Tod is inspired by impressionistic imagery. He combines aspects of pixelation, colour and video noise poetically, using intertitles from the diary of Jean-Paul Sartre. The film is constructed and conceptualized through personal observations sharing intimate memory fragments. In *Mobile-Mentary – Mobile Documentaries in the Mediascape*, Schleser refers to *Italian Love Meetings* and *Why Didn't Anybody Tell Me It Would Become This Bad in Afghanistan*, which also express these characteristics. *Italian Love Meetings* by Marcello Mencarini and Barbara Seghezzi mainly makes use of close-up shots and relies predominantly on vox pops related to the title of the mobile-mentary. Dutch avant-garde filmmaker Cyrus Frisch tells the fictional story through the eyes of a Dutch Afghanistan veteran dealing with the tensions and conflicts of immigrants within Dutch society in *Why Didn't Anybody Tell Me It Would Become This Bad in Afghanistan* (Schleser 2011, 44 and 45). In sync with the presented argument, these two mobile films rely on the intimate qualities and the significance of personal filmmaking in mobile and now smartphone filmmaking. It is key to point out that these filmmakers, as

Figure 2.5 Still image: *SMS Sugar Man* (Kaganof 2007).

much as the second wave of mobile and smartphone filmmakers in the years 2008–12, had access to multitudes of other cameras but chose to use the mobile phone considering its aesthetic qualities. Cyrus Frisch as much as Mark America or the French filmmakers (discussed subsequently) further developed their creative backgrounds from independent film and new media art into mobile filmmaking.

Mark Amerika is an internet artist and VJ who filmed *Immobilité* in Cornwall in the UK in 2007/2008 (Figure 2.6) (Amerika 2008). He describes the feature-length mobile phone art film as 'mashing up the language of "foreign films" with landscape painting and literary metafiction' (Amerika 2008). While the film is self-reflective on filmmaking and cinema, forecasting its future, the backstory reveals 'a future world where the dream of living in utopia can only be sustained by a nomadic tribe of artists and intellectuals' (Amerika 2008). Mark Amerika filmed *Immobilité* with a Nokia N95. While the iPhone was released in 2007, it becomes only of interest for filmmakers with the iPhone 3GS model in 2009, which could film at a VGA resolution (640 × 480). With the launch of the iPhone 4S in 2011 which was capable of 1080p video, the industry started to recognize and utilize the smartphone for filmmaking and broadcast journalism purposes as 1080p (1920 × 1080) was the television broadcast standard at that time.

At the Pocket Film Festival in Paris, smartphone feature-length films by Jean-Claude Taki, Lionel Soukaz, Alain Fleischer and Rachid Djaïdani were screened (Figure 2.7) (Labourdette 2010). The filmmaker's own backgrounds influenced their mobile feature film productions that transcend the domain of experimental and documentary filmmaking. In *La Ligne Brune*, Rachid Djaïdani is portraying the pregnancy of his wife, coupled with a commentary on his perception of contemporary social issues in France through a critical lens. Jean-Claude Taki's *Sotchi 255* explores the mobile aesthetics and notions of memory in poetic and essayistic form. Memory provides the key to read the film *Sotchi 255*, while his short films are colourful fusions of light, flowers and the female body. The mobile film *Le Monde vu par mes jouets* by Alain Fleischer illustrates this mobile phone aesthetic. Similarly his earlier mobile films are *flâneur* like, an exploration of cityscapes through the lens of the mobile camera phone in almost one continuous take. When writing about the Pocket Film Festival as a pivotal contribution to contemporary French cinema, one can point at another example to underline the innovative position France takes in mobile filmmaking. Godard's 2010 project *Film Socialism* incorporates mobile phone video. This is not surprising as Godard, as a video pioneer, commenced working with video technology in 1968 for TV productions in the projects *6 fois 2* and *France tour deux enfants*. At the time of writing this book, Samsung introduced 8K smartphone video capacity. More significantly than mega-pixels is the freedom to explore and experiment. Here one can not only draw parallels to the Dziga Vertov Group by the French New Wave filmmakers, but one could also consider the Oberhausen Manifesto or Doga Filmmaking schools. The key is that filmmakers defined their understanding of film collectively and positioned this not entirely into the service of the entertainment industry.

Figure 2.6 Production still: Mark Amerika's *Immobilité*, a feature-length mobile phone art film.

12. Les invités - Festival Pocket Films 2010

Les invités

Confier des caméras de poche à des plasticiens, écrivains et cinéastes est l'une des missions du festival. Rachid Djaïdani et Lionel Soukaz ont été conviés à ouvrir leur pratique à la phone cam. Jean-Claude Taki et Alain Fleischer, praticiens aguerris mais toujours enthousiastes, présentent leurs premiers longs métrages pocket ainsi que des rétrospectives de leurs courts. Les procédés novateurs du web ne sont pas oubliés avec un aperçu du travail du vidéoblogger Max Schleser.

Jean-Claude Taki

Sotchi 255 { sa 19 juin 16h15 Salle 300 }
Rétrospective des courts métrages { sa 19 juin 18h30 Salle 300 }

Les films de Jean-Claude Taki, évoluant entre fiction et documentaire, travaillant l'image des pays de l'Est, sont des voyages dans la mémoire et des quêtes intimistes. Le réalisateur vient présenter ses courts métrages ainsi que son long métrage inédit *Sotchi 255* sur la quête d'une amie russe disparue. Dans le Grand Foyer, l'installation "Autoportrait 365" est le résultat d'une année d'autoportraits.

➤ Projection de *Sotchi 255* en avant-première de la sélection en compétition au FIDMarseille 2010

Lionel Soukaz

Journal à deux mains { sa 19 juin 20h30 Salle 100 }

Le cinéma de Lionel Soukaz, d'une grande force plastique et émotionnelle, se situe à la confluence du cinéma de l'intime et du militantisme. *Journal à deux mains* a été tourné spécialement pour le festival Pocket Films, en collaboration avec le rappeur et écrivain Tripak. "Avant, le super 8 était un outil idéal pour filmer et montrer les sujets tabous comme l'homosexualité : hors commerce, hors censure, on avait une véritable liberté. C'est sans aucun doute ce qu'on retrouve avec le portable aujourd'hui."

13. Les invités - Festival Pocket Films 2010

Alain Fleischer

Rétrospective des courts métrages { di 20 juin 14h30 Salle 300 }
Le Monde vu par mes jouets { di 20 juin 16h30 Salle 300 }

L'œuvre d'Alain Fleischer s'étend de la littérature au cinéma, en passant par la photographie et les installations. Il est le fondateur et directeur de l'école du Fresnoy qui dispense un enseignement sur les nouvelles technologies aux diplômés d'écoles d'art. Outre ses pocket films, Alain Fleischer présente le "work in progress" de sa dernière œuvre, *Le Monde vu par mes jouets*. Dans cet essai, la caméra est "embarquée" sur des mobiles, en particulier des jouets d'enfants.

Rachid Djaïdani

La Ligne brune { di 20 juin 15h30 Salle 100 }

Écrivain ("BoumKoeur", "Mon Nerf", "Viscéral"), acteur (*Ma 6-T va crack-er*), réalisateur, Rachid Djaïdani nous présente en exclusivité son deuxième long métrage. "*La Ligne brune* est un film construit sur les neuf mois enveloppant la grossesse de Sabrina, ma femme, et qui, à fleur de peau, fera éclore sur le flow de sa réflexion son bonheur : devenir mère. Être père ? Je n'ai pas fait d'école et mon instinct animal m'a largué depuis Cro-Magnon. (...) *La Ligne brune*, une échographie filmique singulière, un hymne à l'amour. »

Max Schleser

Max with a Keitai { sa 19 juin 15h00 Salle 100 }

Max Schleser se définit comme un mobile-filmmaker. Il est le fondateur en Angleterre du projet Filmobile, réunissant les industries de la téléphonie mobile, des artistes et des cinéastes. Il s'intéresse particulièrement aux formes alternatives de tournage en relation avec les technologiques innovantes. Il a développé le concept de mobile-mentary, dont *Max with a Keitai*, tiré de son expérience de vidéoblogger au Japon en 2006, est un exemple.

Figure 2.7 Festival programme: Pocket Film Festival 2010. Mobile filmmaking pioneers in France.

2.2 An original film form: Mobile, pocket and smartphone filmmaking

The early work by the pioneers as much as the more recent projects featured in the International Mobile Innovation Screenings demonstrate the key characteristics specific to pocket and smartphone filmmaking:

- More personal and more authentic stories are surfacing.
- Greater access to storytelling.
- Community engagement strategies.
- Smartphone filmmaking as a global phenomenon with festivals in Asia and Africa.

This chapter defines smartphone filmmaking with its distinctive categories and modes. Following the well-established framework of Bill Nichols's (2001 and 2010) documentary modes or more recently the i-docs (interactive documentary) modes (Aston, Gaudenzi and Rose 2017), this chapter introduces the poetic and experimental mode, participatory and engagement mode, smartphone native mode, 'Red carpet in your pocket' mode and conversational mode.

As Nichols, as well as Aston et al., pointed out, the characteristics of a given mode function as a 'dominant in a given film . . . but they do not dictate or determine every aspect of its organisation' (Nichols 2001, 100). This study will define these five modes through the mobile, pocket and smartphone films, which were selected and presented at the International Mobile Innovation Screenings between 2011 and 2019, and Section 6.4 outlines these modes across several projects in media, arts and design.

While there are many smartphone films available online, MINA (Mobile Innovation Network and Association) presents these films in a curated and reviewed programme (Figure 2.8). A dedicated screening for mobile, pocket and smartphone films provides a means and forum for a dialogue on this subject (Figure 2.9):

- Smartphone films can be presented on the big screen. When the question is asked which smartphone model is the best model to use, the answer should be the one in your pocket.
- One can only talk about smartphone films in relation to other smartphone films (when discussing a narrative genre film, one compares this to a film within its genre, or when talking about a body of artwork or a design artefact this is related to the same field in art and design).
- Creating a community of practice with filmmakers who are interested in exploring the creative potential. Smartphone film festivals and screenings play a major part in showcasing and celebrating new trends and directions. The smartphone filmmaking discourse is created by independent filmmakers and Creative Arts researchers outside the studio system.
- Smartphone filmmaking is an international phenomenon, with distinctive festivals in Asia and Africa, demonstrating opportunities for more localized filmmaking to emerge.

2.3 MINA (Mobile Innovation Network and Association)

MINA provides a platform for smartphone film screenings and discussions. In a recent *Media NZ* journal article, Schleser outlined this digital community and credited the collaborations across countries, time zones and institutions (Schleser 2018). Furthermore, this book surveys filmmakers that have explored distinctive approaches to smartphone filmmaking and have given a voice to individuals and communities not represented in the mainstream media. Smartphone filmmaking provides a means to work with people in co-creative and collaborative ways. Likewise, this book hopes to capture the community of smartphone filmmaking on an international scale and showcase projects beyond the mainstream.

This book showcases smartphone filmmaking's capacity for innovation and transformational creativity. MINA sees itself as connecting communities and the Creative Industries. MINA is not only about showcases but celebrates smartphone filmmaking, and its mission is to engage communities in smartphone filmmaking. In order to make the

Figure 2.8 Third International Mobile Innovation Screening (Auckland, New Zealand, Aotearoa, 2013) featuring screening by MINA's partner festivals: iPhone Film Festival (United States), SeSiFF (South Korea), CinePhone (Spain), International Mobile Film Festival (United States), Ohrenblick mal and Mobil Streifen (Germany) and Mobile Film Festival (Macedonia) (MINA 2013).

Figure 2.9 MINA panel discussion following the eighth International Mobile Innovation Screening (MINA 2018).

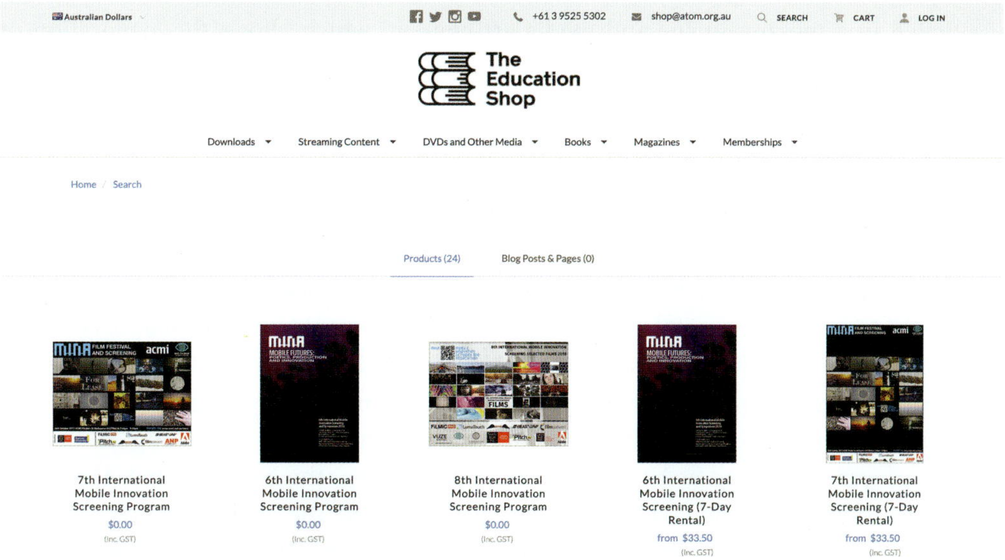

Figure 2.10 ATOM education shop (ATOM 2019).

screening reel accessible, it is now distributed on demand via the ATOM (Australian Teachers of Media) education shop[5] and the screening programmes on the MINA website[6](Figure 2.10).

2.4 Curating the International Mobile Innovation Screening

The story of this book begins with the FIL*M*OBILE[7] screening and exhibition in London in 2008. Max Schleser curated the events as part of his PhD research at the University of Westminster in the Centre for Research and Education in Arts and Media working with Michael Mazière, curator in the London Gallery West. In 2008 only three English language feature-length mobile phone films existed, and the main objective was to bring these mobile films together in one screening to start a conversation with the new film look and feel of mobile filmmaking. The screening in the Lumiere Cinema as much as the exhibition in the London Gallery West was curated from a filmmaker's position and a Creative Arts research perspective. By means of bringing mobile filmmaking and moving-image arts practices into a dedicated space, an analysis with a focus on mobile aesthetics and characteristics could be established. In 2008 mobile filmmaking was a future vision that was shared by filmmakers and audiences alike. The Lumiere Cinema in Regent Street as the birthplace of British Film with the first-ever screening of the Lumière Films in the UK in 1896 was an excellent venue for making a bold statement about some of the changes in filmmaking that have now been realized and recognized in the industry with smartphone films being screened at Cannes and Berlinale[8] and receiving blockbuster distribution deals as in the case of Steven Soderbergh's *Unsane*. *Smartphone Filmmaking: Theory and Practice* as much as MINA's curatorial practice is situated in a moving-image and media art context, which allows the exploration of aesthetics, unconventional practices and new formations. While this book explores the art and business of smartphone filmmaking, it is positioned in the tradition of the moving-image arts, because mobile and smartphone filmmaking emerged out of this space. While storytelling is significant for certain projects and especially community engagement, it is not the only or determining factor to recognize.

The MINA screenings in the last nine years (2010–19) included a discussion and presentation of the works by smartphone filmmakers, artists, designers and makers (Figure 2.11). As the name of the annual screening indicates, *Mobile* is also understood as a conceptual set-up. Over the last years, MINA moved screening locations every two to four years in order to make new connections with filmmakers and audiences alike. The previous nine editions of the International Mobile Innovation Screening & Festival screenings were presented at Ngā Taonga Sound & Vision, the New Zealand Archive of Film, Television and Sound (Wellington, NZ, 2011 to 2018), the Australian Centre for the Moving Image (ACMI) (Melbourne 2017), AUT University (Auckland 2013 and 2014), RMIT University (Melbourne 2015), Lido Cinemas (Melbourne 2016), Swinburne University of Technology (Melbourne 2018) and IMSC – International Mobile Storytelling Congress at the University of Nottingham Ningbo China (Ningbo 2020) (Figure 2.12).

Writing in *Rethinking Curating*, Graham and Cook point out that artists might be best placed to curate an emergent medium ('because they work with it, they know most

Figure 2.11 MINA panel discussion following the eighth International Mobile Innovation Screening (from left to right: Martin Koszolko [mobile musician/music producer], Ajax McKerral [Hoopy], Angela Blake [see Section 3.4, SF3], Nicholas Hansen [filmmaker], Max Schleser [MINA screening director], Eamon Wyss [Hoopy], Patrick Kelly [filmmaker], Rachel Chen [emerging filmmaker], Chloe Marie Villasanta Samillano [emerging filmmaker], Mikee [VR Cinema] [MINA 2018]).

about it') (Graham and Cook 2010, 254). Graham and Cook refer to curator Steve Dietz who states:

> new media art is like other contemporary art, but it also has particular characteristics that distinguish it from contemporary art and by extension from the system involved in the production, exhibition, interpretation, and dissemination of contemporary art – the realm of curators.
>
> (2010, 1)

This described relationship and its similarities and simultaneous differences are in sync with the interactions of smartphone filmmaking with the contemporary film industry. The interviews by filmmakers illustrate this view in terms of production and the interviews with smartphone film festival directors relate to the notion of exhibition. The in the quote described 'particular characteristics' are defined through the Smartphone Filmmaking modes presented in Sections 2.6 and 6.4 in this book. The dissemination of smartphone filmmaking is discussed in several chapters throughout this book, through the smartphone film festivals, which provide an alternative forum to industry distributors, and in the context of Mobile XR pointing towards new forms and systems of dissemination and exhibition.

In the first years, 2011–14, the screening reel developed the argument that mobile, smartphone and pocket cameras can produce work for cinematic projection. The curatorial

Figure 2.12 MINA panel discussion at ACMI featuring Hilary Davis (Social Innovation Research Institute, Swinburne University), Jason Van Genderen (Section 5.5), Tania Sheward (moderator and digital storyteller), Adrian Jeffs (award-winning smartphone filmmaker) and Max Schleser (MINA screening director and smartphone filmmaker) (MINA 2018).[9]

approach referenced the tradition of avant-garde filmmakers, the explorations of new approaches and creative strategies, which defined the aesthetics of mobile media art (Baker, Schleser and Molga 2009). Smartphone filmmaking's specificity is a key curatorial element in the screening. The selection focused on mobile and smartphone films which demonstrate personal and intimate storytelling, capturing ephemeral moments, semi-permanent situations and interventions in everyday life. The affordances of the small scale and the always ready to record parameter allow more creative experimentation and also opportunities for social engagement through voices that are not heard in the mainstream. The MINA screenings showcase mobile, smartphone and mobile film screenings that deal with these or interrelated creative explorations. MINA is curated with a bold creative vision to screen mobile and smartphone films, video and moving-image works that might not be screened on the big screen otherwise. In the last years, MINA also screened works that can be situated in the domain of mobile social media, Korsakow filmmaking,[10] drone cinematography, AR and VR.

While MINA operates within the format of a cinema screening, alternative approaches were tested at the Fringe Festival in Wellington (2013) including a public projection using

Figure 2.13 MINA @ Fringe Festival. Public projection on Cuba Street in Wellington, New Zealand (MINA 2013).

a projector and a generator in a parking lot on Wellington's Cuba Street and screenings during community workshops (Figure 2.13). Following a four-day screening programme in Ngā Taonga Sound & Vision, in 2014, the decision was made to work towards a 90-minute screening reel, allowing for the curated screening to be viewed by a larger audience. Therefore, the focus on short mobile, smartphone and mobile films (under 7 minutes) was established in 2015. The screening reel was also featured at a number of other festivals internationally including Le Jour Le Plus Court, Strasbourg, France, December 2013; Telefon Art at the European Film Festival Szczecin, Poland, December 2016; *Small Screen Art, Filmwinter*, Stuttgart, Germany, January 2014, and the Thirteenth Festival International de la Imagen, Columbia, May 2014, among others, as well as conferences (i.e. Visible Evidence XXII Toronto, Canada, August 2015 or Expanding Documentary Conference, AUT University, Auckland, December 2011).

While some smartphone films might be available online on the filmmakers' YouTube or Vimeo channels, the curated screening brings together smartphone filmmakers from around the world. As diverse as the nationalities are the backgrounds of the smartphone filmmakers themselves. Each year the screening programme is curated to celebrate creativity and diversity by means of choosing filmmakers from various backgrounds, including but not limited to documentary filmmaking, video art, new media art, creative technologies, community-engaged projects, experimental filmmaking as well as creatives and video producers from advertising and marketing. Every year the screening combines smartphone filmmakers who had previously screened their work at MINA, smartphone

Figure 2.14 MINA smartphone filmmaking workshop by Max Schleser. St Kilda Film Festival, Melbourne, Australia (SKFF 2018).

filmmakers new to screening at MINA or showcasing first-time smartphone filmmakers. Most smartphone filmmakers who were part of the MINA screening joined the loosely connected community of smartphone filmmaking online. There are several Facebook groups and Twitter conversations, which form a community of practice (Wenger 1999). In addition to the screening, MINA runs workshops to engage and educate audiences, community groups and various industry sectors about smartphone filmmaking.[11] For the MINA screening reel as much as the workshop, smartphone filmmaking is about accessibility and creative experimentation. The workshops and screenings highlight mobile filmmaking's capability for various interactions with communities and Creative Industries. Smartphone filmmaking talks took place at festivals such as Pause Fest and St Kilda Film Festival (both Melbourne, Australia), and at Doc Edge Incubator (Auckland, New Zealand) (Figure 2.14). A number of community-engaged projects for NGOs and community groups are part of MINA's work beyond the screening and showcase activities. As a network, MINA brings together various creatives and connects Creative Industries to communities.

The question concerning smartphone filmmaking and its contribution to the theory and practice of filmmaking is no longer limited to only the discussions in the smartphone filmmaking community. Here we can refer back to new media, which

> fundamentally challenge notions of authorship, and the network model of new media is not the broadcast model of one to many, but the many-to-many approach of peer-to-peer networking. The roles of artist, audience and participant are blurring: the participants sometimes *are* the audience. An audience of peers is excellent for pushing the expertise and critical level within a field.
>
> (Graham and Cook 2010, 181)

Very much as the Dziga Vertov Group of the French New Wave filmmakers shot on location with 16-mm cameras, used improvisation and developed new ways of editing (i.e. jump cuts), smartphone filmmakers explore and refine the aesthetics of mobile media art. In a similar way as Dogma filmmakers developed a new approach to filmmaking as a reaction against the studio system, smartphone filmmakers create an alternative space. Several manifestos and movements, such as the Vertical Cinema manifesto or #Filmbreaker[12] movement, appeared in the last years. As much as MINA in Australasia other smartphone film festival directors and organizers are also energetic smartphone filmmakers, which emphasizes the blurring of roles and community approach – for example, Benoît Labourdette and Simon Horrocks create a space for filmmakers to network, get together and share their work. They influence smartphone filmmaking not only through their creative film practice but also through building a community. As the film industry did not recognize mobile filmmaking for about a decade, it operated very much situated in an independent space and in grassroots and bottom-up approach, opposing to the hierarchical structures in the film and TV industry. While smartphone's technical capability is continuously increasing, more and more apps become available to cover the entire production process for on-the-go capabilities, and new add-ons and accessories (such as Gimbals and microphones) meet technical industry standards required for certain professions (e.g. mobile journalists or mojos).

MINA is an audience of peers, and every year it provides a reference point for the trends and directions in smartphone filmmaking. The MINA judges reflect the diverse community with a range of different backgrounds internationally. Rather than asking prominent figures from the TV and film world, the MINA committee is composed out of smartphone filmmakers, artists working with mobile media and smartphones as well as researchers and educators, who are constantly exploring new ideas and challenge the status quo. This also means that MINA is working within the Creative Arts research environment with reference to quality assurance and peer review, which is applied to the selection process each year. Utilizing the social media avenues, MINA Facebook group[13] and page as well as the FilmFreeway[14] platform, made this process more streamlined. The 2011 edition operated via paper-based submission forms and DVDs that were mailed from around the world. Within the last nine years, this process shifted into the cloud. In the 'Mobile XR' section of this book, some opportunities for new dissemination systems are discussed, which are currently being explored. The MINA committee is reflective of the smartphone filmmaking community, which set this movement into motion and continues to expand internationally.

2.5 International Mobile Innovation Screening, 2011–18

Table 2.1 provides a taxonomy according to the proposed modes in a chronological order as presented at the International Mobile Innovation Screenings from 2011 to 2018. These films were chosen as they characterise these modes explicitly. The cited mobile and smartphone films are discussed in the following sections.

Table 2.1 International Mobile Innovation Screening, 2011–18

Year	Mode	Mobile, pocket and smartphone film
Screening 2011 (Figure 2.15)		
	Poetic and experimental mode	
		July (Joao Krefer)
		Memory Cathedral (Dean Keep)
		Novo Ano (Louise Botkay Courcier)
		Moscow Diaries (Adam Kossoff)
	Participatory and engagement mode	
		Expose Yourself (Anders Weberg)
		24Frames 24 Hours (Max Schleser)
		Reel Health (Joanna Ong)
	Smartphone native mode	
		N/A as Apple app and/or Google Play store launched that year
	Conversational mode	
		Yours & Mine (Anne Massoni)
		Emphemeral Mobile Media VJing (Camille Baker)
		Switched Eyes (Caroline Bernard, Michiko Tsuda and Damien Guichard)
	'Red carpet in your pocket' mode	
		God in My Pocket (Arnault Labaronne)
Screening 2012 (Figure 2.16)		
	Poetic and experimental mode	
		Tamarindo Costa Rica (Felipe Cardona)
		Sinfulness (Anders Weberg)
		Poster (Benoît Labourdette)
	Participatory and engagement mode	
		Massey Time Capsule (Max Schleser)
	Smartphone native mode	

Year	Mode	Mobile, pocket and smartphone film
		Day at Amagansett Beach (David Scott Leibowitz)
		5#CALLS (Giuliano Chiaradia)
	Conversational mode	
		Homemade Space Craft (Luke Geissbühler)
	'Red carpet in your pocket' mode	
		Finn (Visual Cooks)
		The Editor (Chris Nong)
Screening 2013 (Figure 2.17)		
	Poetic and experimental mode	
		The 57' (Leo Berkeley)
		Mobilearte (Gerda Cammaer)
		Hydroscope (Kaihei HASE)
		Kolacze (Wilhelm Jerusalem)
	Participatory and engagement mode	
		Yours and Mine 2 (Anne Massoni)
	Smartphone native mode	
		About 7am: The First Quarter (Adrian Miles)
		Midtown (Max Schleser)
	'Red carpet in your pocket' mode	
		Damp (Jake Ngawaka)
	Conversational mode	
		Two Stories (Zaher Omareen)
		Digital Trust Hike (Paul Taylor)

(*Continued*)

Table 2.1 (Continued)

Year	Mode	Mobile, pocket and smartphone film
Screening 2014 (Figure 2.18)		
	Poetic and experimental mode	*Absent* *Numb* (Anders Weberg)
	Participatory and engagement mode	
		Study on Persistence (Kefer)
		Here and There (Felip Cardona)
	Smartphone native mode	
		David Hockney at the Met (David Scott Leibowitz)
	'Red carpet in your pocket' mode	
		Calling (Xavier Satorra)
	Conversational mode	
		Tofu Man (Andrew Robb)
Screening 2015 (Figure 2.19)		
	Poetic and experimental mode	
		Us Stop (Julien Tatham)
		Diaorama (Patrick Kelly)
	Participatory and engagement mode	*2 min 2 hours* (Gwendoline Rippe)
	Smartphone native mode	
		In the Heart of the Kaleidoscope (Vanessa Vox)
		Pacific Colours (Max Schleser)
	'Red carpet in your pocket' mode	
		Dog Tail (Swathy Deepak)
		The Life and Death of the iPhone (Paul Trillo)
	Conversational mode	
		Gasp (Anna Jones)
		In Response We Closed Flinders (Alex Dick)

Year	Mode	Mobile, pocket and smartphone film
Screening 2016 (Figure 2.20)		
	Poetic and experimental mode	
		Dark Waves (Sven Dreesbach)
	Participatory and engagement mode	
		Give Me Faces and Streets! (Patrick Kelly)
		Falling On Cement (Leo Berkeley)
	Smartphone native mode	*On the Move* (Laurent Antonczak, Felipe Cardona, Max Schleser and Daniel Wagner)
	'Red carpet in your pocket' mode	
		Bubbles Don't Lie (Stepan Etrych)
	Conversational mode	
		Connectivity – dis –Connectivity (Anagha Saggar)
		Emergency Call – Appel d'urgence (Brice Veneziano)
Screening 2017 (Figure 2.21)		
	Poetic and experimental mode	
		How Was Your Birthday? (Mohammed Zaouche)
		The Big City (Evan Luchkow)
	Participatory and engagement mode	
		#Selfiesuit Documentation (Paul Taylor)
	Smartphone native mode	
		Dreaming (Vanessa Vox)
	'Red carpet in your pocket' mode	*You're a guy!* (Sylvain Certain)
	Conversational mode	
		Unseen Memories (Manpreet Singh)
Screening 2018 (Figure 2.22)		
	Poetic and experimental mode	

(*Continued*)

Table 2.1 (Continued)

Year	Mode	Mobile, pocket and smartphone film
		A Guide to Breathing Underwater (Raven Jackson) *White Point* (Annette Philo)
	Participatory and engagement mode	
		#Selfiesuit_departure_V2 (Paul Taylor)
	Smartphone native mode	
		Ice Swimming (Robb Montgomery)
	'Red carpet in your pocket' mode	*She Rose* (Malwina Wodzicka)
	Conversational mode	
		Kuujjuaq (Sammy Gadbois) *What's with the Nails?* (Patrick Kelly)
Screening 2019 (Figure 2.23)		
	Poetic and experimental mode	
		Night Sail (Gerda Cammaer) *Irradiate* (Paul London)
	Participatory and engagement mode	
		Race against Racism (James Hyams) *Gray Matter AR* (Karen Vanderborght)
	Smartphone native mode	
		Uncommon (Faraaz Fakhri)
	'Red carpet in your pocket' mode	*Friendship* (Galina Fesenko)
	Conversational mode	
		The Reef (Sven Dreesbach)

2.6 Smartphone filmmaking modes

Mobile, smartphone and mobile filmmaking is defined as an original film form through the following five modes:

1) Poetic and experimental mode: Twenty-first-century experimental film and time-based media, which is not only exhibited in the gallery or museums context but is also situated online and in other semi-permanent and ephemeral forms and formats.
2) Participatory and engagement mode: Co-creation, collaboration and community-engaged media texts relate closely to Nichols's participatory documentary mode (Nichols 2001) but expand this linear and narrative form by engaging audiences actively as participants and co-creators into the creative process and smartphone film production. The creative process in the participatory and engagement mode in smartphone filmmaking shifts from script to screen to shoot and share. The dissemination element is no longer only part of the final project phase but allows participants and co-producers to contribute to the production.
3) Smartphone native mode: Filmed and edited on smartphones via apps. Smartphone native video terminology is adapted from social media, which references native videos as 'videos that are uploaded to or created on social networks and played in-feed, as opposed to links to videos hosted on other sites' (Adweek 2015). In the context of smartphone filmmaking, this means videos are filmed, edited and uploaded from mobile devices and smartphones. As opposed to desktop editing in a studio, the specificity of mobile devices affords novel approaches to editing and storytelling while being out and about. Some smartphone film festivals, such as SF3 or FiLMiC Fest, refer to these emerging forms as 'made on mobile'.
4) Conversational mode: As mobile devices characterize the fusion of communication and lens-based media, this conversational mode is characterized by opening up a dialogue through storytelling. The accessibility of the smartphone affords more voices coming to the screens and thus democratizing storytelling.
5) 'Red carpet in your pocket': Narrative films which use Hollywood conventions to produce smartphone short or feature films. The terminology is adapted from Karl Bardosh and the International Mobile Film Festival.

The above order of the modes reflects their significance for this research and mapping smartphone filmmaking internationally. The poetic and experimental mode is emerging out of alternative filmmaking traditions and can be linked to constructivist, abstract, structuralist or flicker filmmaking approaches and styles. The filmmakers in this canon do not necessarily work with scripts but foreground a concept and very much the process in filmmaking, including self-reflective filmmaking. The smartphone native mode as much as the conversational mode developed in sync with the proliferation of smartphone, social and

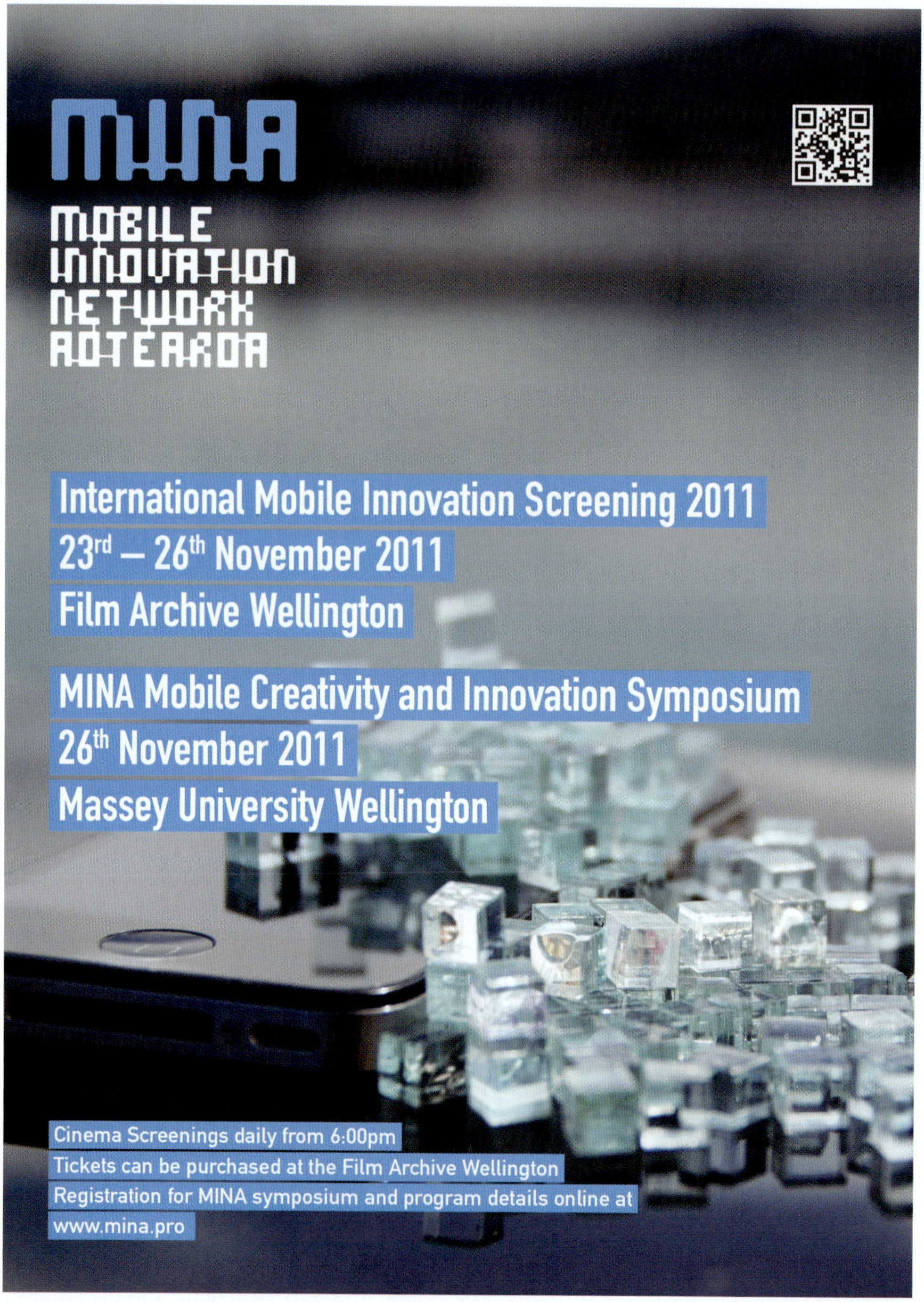

Figure 2.15 MINA Screening Programme 2011 (MINA 2011).

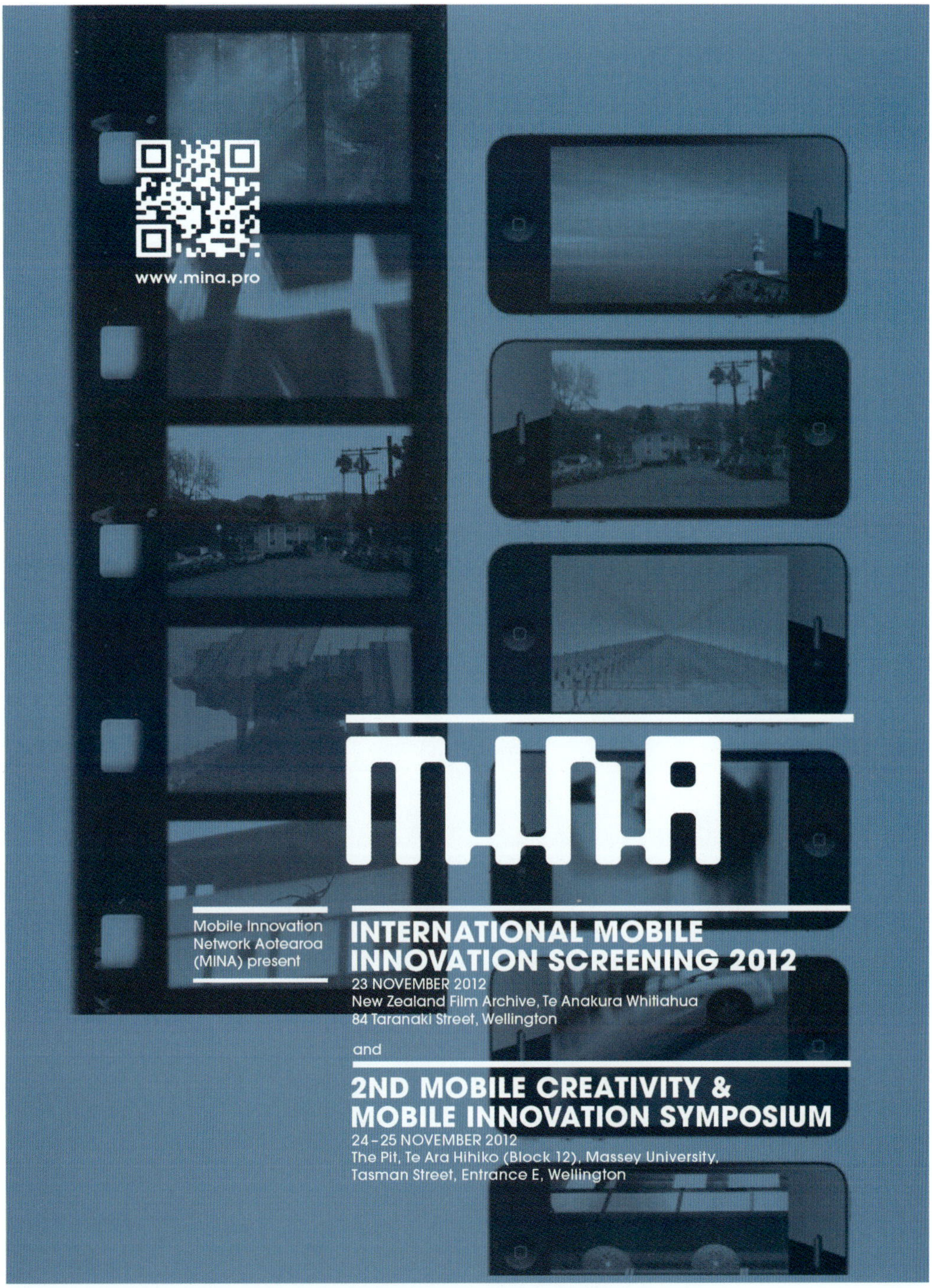

Figure 2.16 MINA Screening Programme 2012 (MINA 2012).

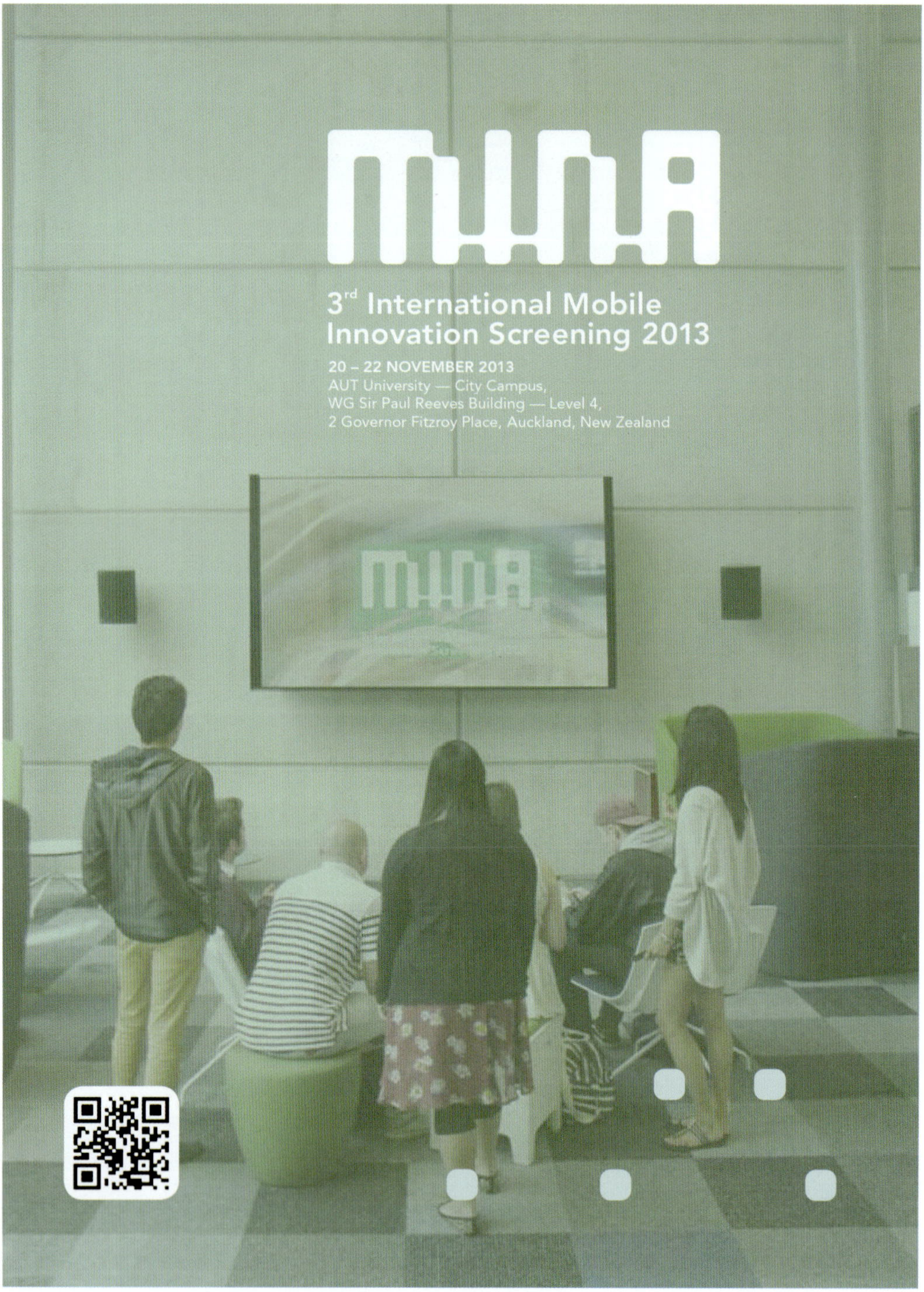

Figure 2.17 MINA Screening Programme 2013 (MINA 2013).

Figure 2.18 MINA Screening Programme 2014 (MINA 2014).

Figure 2.19 MINA Screening Programme 2015 (MINA 2015).

Figure 2.20 MINA Screening Programme 2016 (MINA 2016).

Figure 2.21 MINA Screening Programme 2017 (MINA 2017).

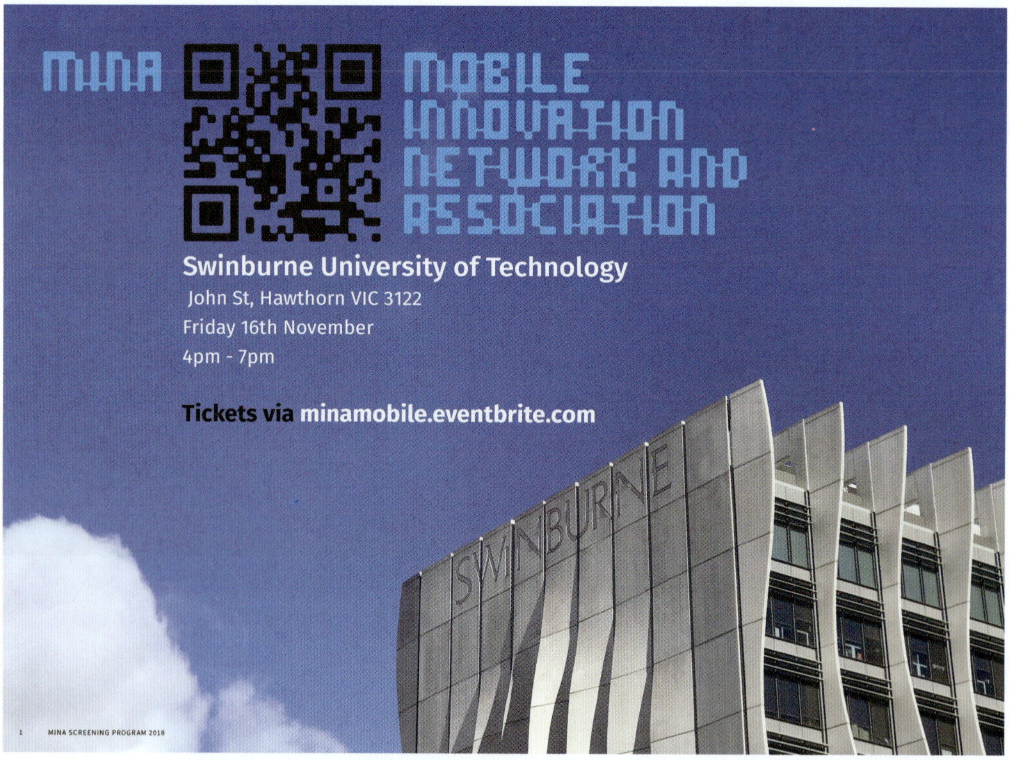

Figure 2.22 MINA Screening Programme 2018 (MINA 2018).

Figure 2.23 MINA Screening Programme 2019 (MINA 2019).

network media as well as video platforms. Out of this ephemeral and online environment, mobile media and mobile cinematic specific aesthetics and formats surface. Later in this book, vertical video is discussed as a smartphone filmmaking specific formation, which has its own video creator personalities that produce prominent work at the intersection of existing industries and new ecologies. The smartphone native mode embraces editing on the go, which means you can edit while being on location, like an artist taking a canvas out of the studio and drawing in the city or landscape (Figure 2.24).

Smartphones are the most accessible cameras and many filmmakers featured in this book, including the author, Max Schleser, conduct smartphone filmmaking workshops. Community engagement is as important to consider as the final outcome for both short and feature-length smartphone films. Schleser's work, as outlined in Section 6.2, demonstrates that smartphone filmmaking can facilitate new collaborative elements and innovative approaches to community outreach and engagement. While new disruptive forms and formats arise out of the smartphone filmmaking domain such as the video-sharing social networking site TikTok, with its microformat of short 15-60 second videos, some filmmakers work in time-based narrative contexts. The 'Red carpet in your pocket' mode remixes existing genres and storytelling formats. The following definitions of the smartphone filmmaking modes are followed by the discussion of smartphone films that screened at the MINA screenings and film festival between 2011 and 2017.

2.6.1 Poetic and experimental mode

The poetic and experimental mode encompasses engagement with aesthetic refinements and simultaneously explores the distinctive visual smartphone look. Furthermore, it embraces experimentation and expands on the established and/or more traditional film practices in framing, editing, imaginative voice-overs and/or fragmented streams of consciousness. The poetic approach magnifies the formal experimentation and interconnects with established formats such as video loops, essay films or/and travel films.

As Rees points out in *A History of Experimental Film and Video* (Rees 2011), there is no coherent formation of experimental film. The following examples were chosen to illustrate the various approaches and processes filmmakers have chosen to experiment with mobile, smartphone and pocket cameras in the poetic and experimental mode. A more personal and authentic engagement with non-narrative and non-linear approaches characterizes the experimental mode. Smartphone filmmaking allows for unrestricted experimentation as we always have our phone always with us. Once an idea crystallizes the camera pen (Astruc 1948, in Corrigan 1999, 158) is ready to sketch out and develop conceptual approaches.

There are a number of parallels that can be drawn between the French tradition of *La Nouvelle Vague* and the contemporary smartphone filmmaking movement. This 1960s movement was a rejection of studio productions working with 16-mm cameras and improvisation in the streets. These characteristics can be used to describe the smartphone filmmaking projects described in this book. The French New Wave developed new ways of filming and consequentially editing that led to the emergence of new techniques such as jump cuts. The world of smartphone filmmaking embraces personal filmmaking and handheld camera. With the dissemination via mobile social media, new formats such as vertical video emerge. In 1948 Alexandre Astruc wrote the article 'The Birth of a New Avant-garde: La Camera Stylo'. In *L'Ecran Français* stating that filmmakers will be able to write with their cameras, he predicted

> a form in which and by which an artist can express his thoughts, however abstract they may be, or translate his obsessions exactly as he does in the contemporary essay or novel. That is why I would like to call this new age of cinema the age of camera-stylo (camera-pen).
>
> (Astruc 1948 in Corrigan 1999, 158)

The argument of the camera-stylo (camera pen) was present in the discussions at the MINA screening at ACMI (Melbourne, Australia, 2017) as much as in the Forum des Images (Paris, France, 2010). Filmmaker Leo Berkeley whose smartphone films screened in the MINA International Mobile Innovation Screening in 2013 (*The 57*'), 2015 (*The Q*) and 2016 (*Falling on Cement*) discussed his essay film *Tram Travels* (2013) linking Astruc's provocations to the genre of the essay film. He quotes Alexandre Astruc promoting the idea that cinema will break away from the concrete demands of the narrative to become a means of writing just as flexible and subtle as written language (Astruc 1948 in Graham and Vincendeau 2009). These notions characterized the Experimental and Emerging Practices panel at ACMI and pointed towards the playfulness and the different experimentation processes. Rather than following a script, filmmakers follow their intuition.

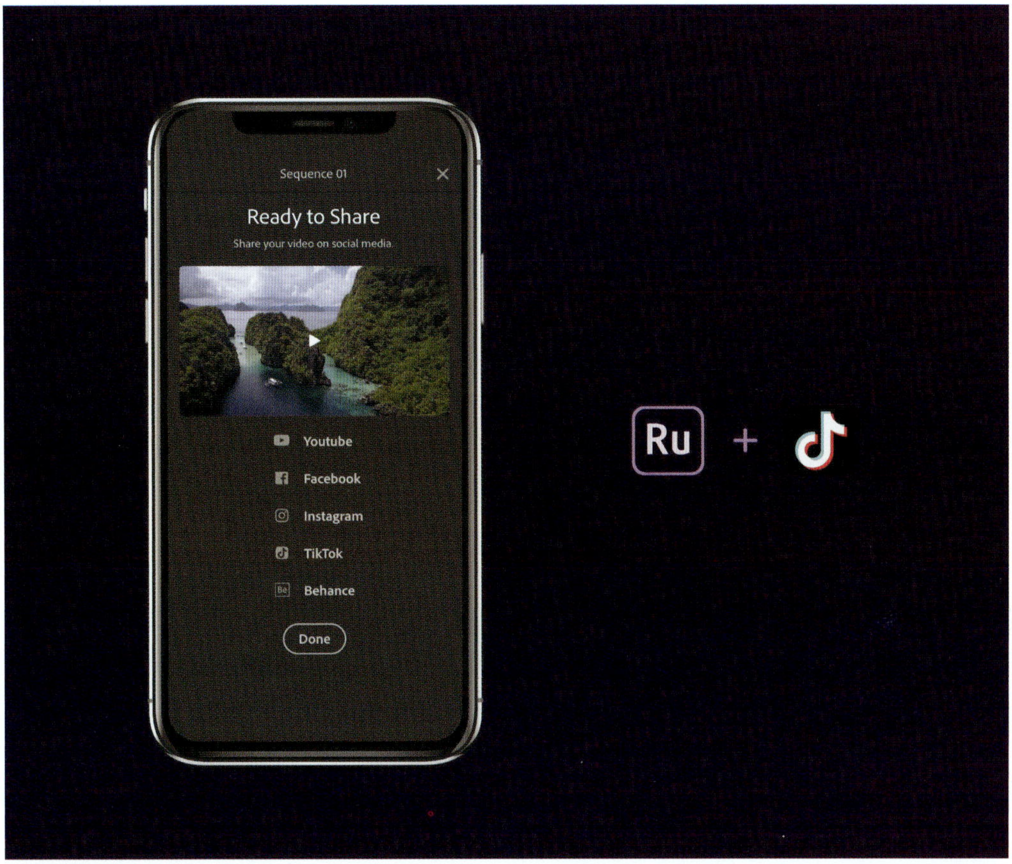

Figure 2.24 Editing on the go with Adobe Rush which integrates into sharing to TikTok (Adobe 2019).

In the same way that an artist starts with a canvas or a designer/engineer with a pencil and paper, a filmmaker can sketch with video. The notion of the everyday as a creative as well as political intervention was prominent in the screening in 2017 as much as in the other MINA screenings. Writing in *Tram Travels: Smartphone Video Production and the Essay Film*, Leo Berkeley supports the formulation of the Keitai aesthetic:

> Schleser's work on the Keitai aesthetic (2014), the aesthetic qualities that emerged as a result involved the ability to convey on screen immediate but fragmented moments of lived experience . . . from a distinctively personal perspective.
>
> (Berkeley 2018, 31)

Despite the increasing technological developments in terms of image quality (3GP in comparison to 2K or 4K video), some characteristics from the Keitai aesthetic, being the intimate and immediate forms of media production, remain present in mobile filmmaking and seem now to shift in the foreground once again.

> This aesthetic [the Keitai aesthetic] emphasises the importance of location . . . . Mobile devices make the mundane interesting, the everyday confronted, providing a new lens for

> viewing the world through a new camera vision. . . . . The sense of intimate connectedness to the message, its subject and the author, has emotional implications in everyday art experience, making one feel special, important and inspired.
>
> (Baker, Schleser and Molga 2009, 119)

The mobile film *Max with a Keitai* was produced in 2006 on location in Japan. Keitai translates from Japanese meaning 'hand-carry' but also has connotations of small, portable, carrying something, form, shape or figure, next to mobile phone (Schleser 2011, 93). The Keitai aesthetic argued for considering a more intimate and immediate connection between the filmmaker and the experience that the work creates. This crystallizes in the filmmaker Max Schleser showing himself in reflections (in trains and mirrors) and through the integration of intertitles in the form of text messages that appeared on his vblog. These text messaging elements were personal comments. As much as text messaging created a new short-form language, the visuals created a new screen media format at that time. Writing in *Mobile Filmmaking* Berry argues that this experience 'moves beyond purely representational aspects of filmmaking techniques and practices into phenomenological domains' (Berry 2017, 312). Furthermore, she makes a link to Schleser's *Connecting through Mobile Autobiographies: Self-Reflexive Mobile Filmmaking, Self-Representation, and Selfies*, highlighting the present tense element in mobile and smartphone filmmaking. The significance of capturing a moment defines the shift from script to screen to a shoot and share approach. While *Max with a Keitai* worked in 2006, these ideas foreshadowed the selfie phenomena which was termed in 2004 but wasn't widespread until around 2012 (Pearsall in Schleser 2014, 150). This is the same year that Facebook acquired Instagram and the mobile media art aesthetics moved into the mainstream. Other mobile smartphone films that emphasize these specific aesthetic qualities and are discussed in the book are Steve Hawley's *Speech Marks*, *Us Stop* by Julien Tatham or Patrick Kelly's *Give Me Faces and Streets!*, among others. Also notable in the mobile feature films are those personal stories and intimate experiences as revealed in *Tangerine*, *Selfie* or *Saudi Runaway*.

The MINA 2011 screening and festival featured Adam Kossoff's *Moscow Diaries* (Figure 2.25). This smartphone film is related to the notion of the essay film, based upon the diary that Walter Benjamin, German cultural commentator, wrote in 1926–7. The film describes Benjamin's efforts to get close to the woman he was in love with, his struggle to get around Moscow and the political debates of the time. Adam Kossoff produced the essay film with the book by Walter Benjamin in one hand and the smartphone in his other, using GPS and the camera app. His use of long shots and voice-over, which are extracts from Walter Benjamin's diary, portrays a personal account fusing time and space.

Felipe Cardona developed an approach to smartphone filmmaking through video loops. He further expands on the soviet montage theories to create novel interrelations between audio and visual tracks. He uses diegetic sounds that create a rhythm for the video. This technique could be described as visual beatboxing or an extension of the metric montage adjusting filmic time to the diegetic sounds captured on location. Simultaneously, his mobile-mentary *Tamarindo Costa Rica MM 22062012* is a travel film capturing his experience of a journey to Tamarindo, Costa Rica.

Figure 2.25 Still image: *Moscow Diaries* (Kossoff 2011).

Gerda Cammaer's *Mobilearte* is positioned in the realm of abstract filmmaking, produced on an iPad. *Mobilearte* is fundamentally about the difference between 'real' and 'experienced' time, especially while travelling. The 'guide track' for this piece is in 'real time' illustrated by the form of the TukTuk ride, complemented by how time was 'experienced' by the filmmaker as fleeting in the city of Maputo, Mozambique. In a similar way, as *Mobilearte* turns an everyday situation into abstraction by creating a rhythm-based on 'moving-images' qualities of form, shape and colour, *Us Stop* by Julien Tatham captures moments of life as people are waiting for the bus in France. Tatham depicts the hidden poetry in the micro details and reveals the extraordinary of the ordinary through the different layers of our reality. He uses layering techniques conceptually not only during editing but also during filming on location. By means of filming the glass walls at a bus stops, Tatham creates visual layers of abstraction, filming reflections on the glass and filming through the glass to create a soft focus.

Following a surreal approach, the Swedish artist and filmmaker Anders Weberg created a homage to the experimental filmmaker Maya Deren (Figure 2.26). The film *Absent VIII a homage to Maya Deren* uses black-and-white cinematography and is a play of light using silhouettes. The reverberation of the audio track underlines the dreamlike state or the stream of unconsciousness as a non-linear approach that also characterizes one of Deren's seminal experimental film, *Meshes of the Afternoon*. Her husband Alexander Hammid filmed Deren on a Bolex 16-mm camera in 1943 capturing 'the inner realities of an individual and the way in which the subconscious will develop, interpret and elaborate an apparently simple and casual incident into a critical emotional experience' (Berger 2010, 303). Next to the aesthetic qualities of personal experiences that resonate in smartphone

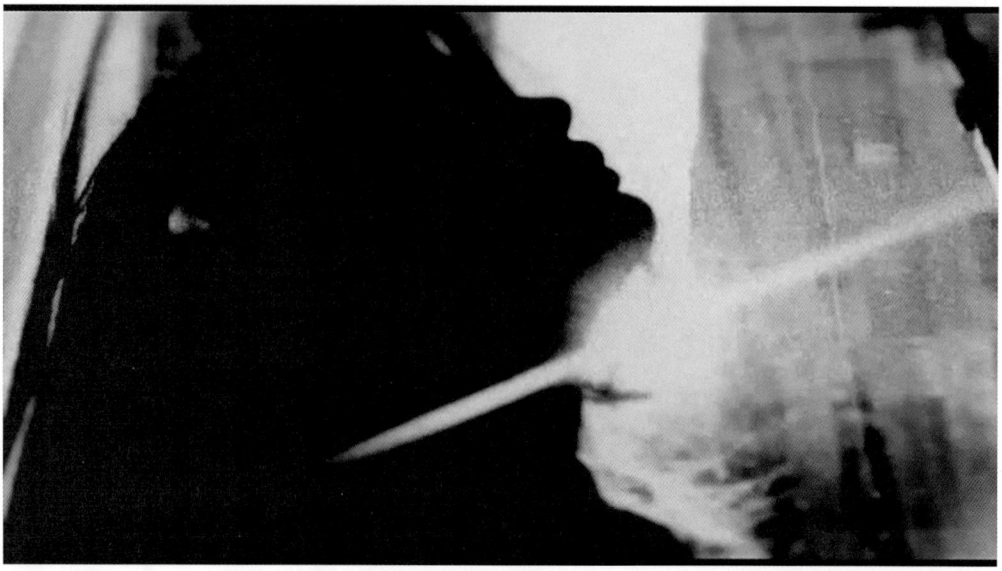

Figure 2.26 Still image: *Absent VIII a homage to Maya Deren* (Weberg 2013).

filmmaking, one can also point towards her understanding of independent and experimental cinema. In her 1965 article *Amateur versus Professional* she points out that the amateur, which originates from the Latin 'lover' and means one 'does something for the love of the things rather than for economic reasons or necessity' (Deren 1965, 45), has

> one great advantage which all professionals envy him, freedom – both artistic and physical . . . . The most important part of your equipment is yourself: your mobile body, your imaginative mind, and your freedom to use both.
>
> (Deren 1965, 45)

This creative freedom is one of the key drivers that resulted in innovative approaches and new perspectives in filmmaking, which the smartphone films in this chapter exemplify.

Evan Luchkow's *The Big City* uses a microscope for cinematography and overlays the microorganisms with the soundscape of Vancouver, Canada (Figure 2.27). His microscopic city film becomes the ultimate extreme close-up. *The Big City* reveals the city from a new perspective as much as Sven Dreesbach's underwater world. Dreesbach captures a lonesome surfer roaming the dark waves of central California's ocean. His work is an exploration of the sea as a rhythm for editing and demonstrates the potential in terms of capturing high-resolution video via FiLMiC Pro app and post-production using low-light and underwater videos with DaVinci Resolve (Figure 2.28).

Wilhelm Jerusalem's *Kolacze* is a music video in which two characters appear in a dreamlike landscape. In this poetic love story, two characters seem to search for each other while never crossing their paths. The storytelling approach uses symbols and metaphors in between a dream world and lived experience; a female character is lying on a field, and a male character seems to wake up from a nap in a shed-like cabin. Both embark on a drift through the woods heading towards a lake. When the male character attempts to cross the river with a boat, the woman wakes up in a cottage. The female character observes herself

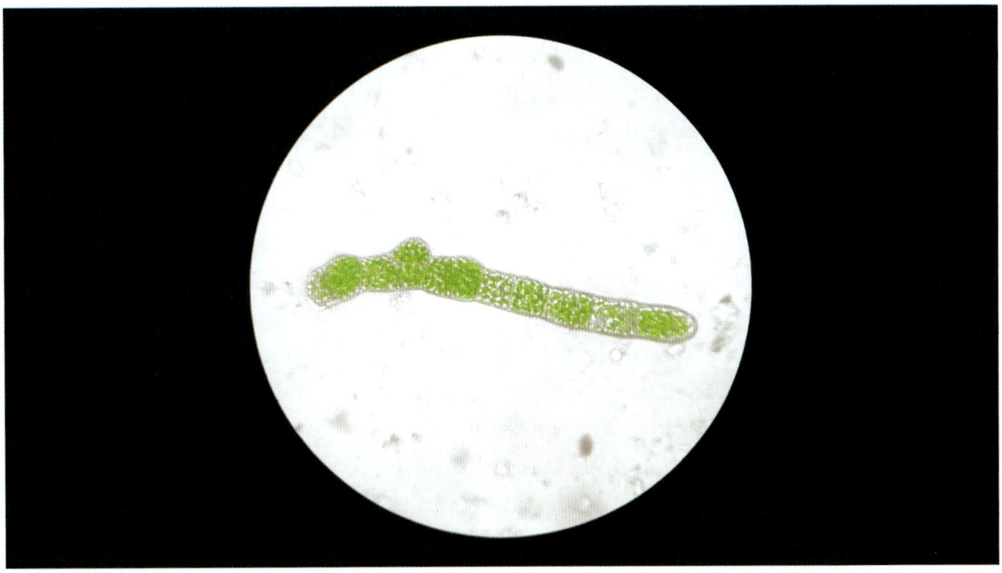

Figure 2.27 Still image: *The Big City* (Luchkow 2018).

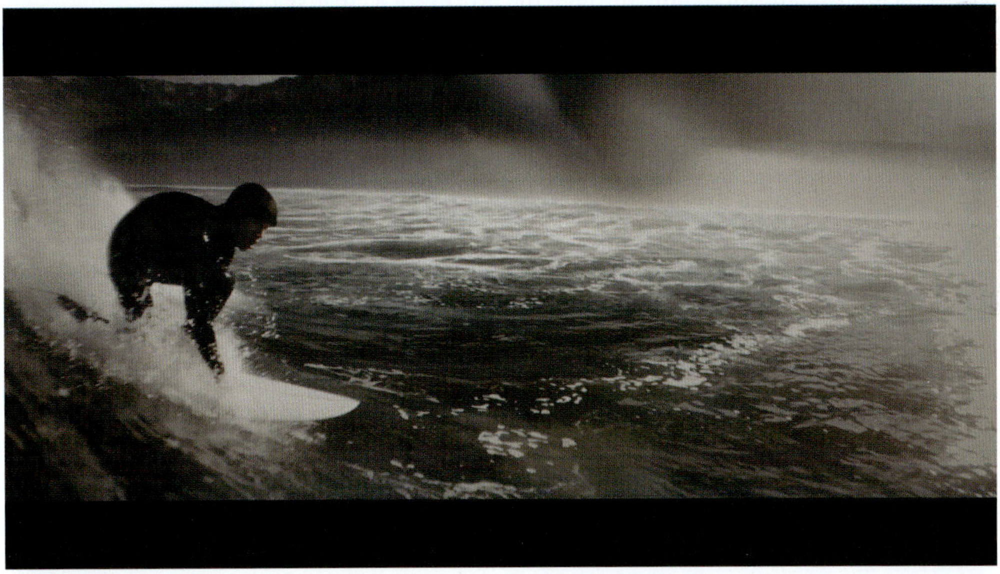

Figure 2.28 Still image: *Dark Waves* (Dreesbach 2016).

in a mirror while the male character encounters a mirror fragment on a tree to find his own reflection as a memory trace. This poetic smartphone film ends with the male character gazing into the landscape that evokes dreamlike connotations.

2.6.2 Participatory and engagement mode

At the core of the participatory and engagement mode, we find collaborative and participatory forms of filmmaking and screen production. Audiences are co-creators and engage with the mobile, pocket and smartphone films online or via face-to-face interaction. While some projects reflect new formations of creativity, others are political or illustrate the importance of community engagement as a form of social innovation. This mode demonstrates that innovation is to be understood as not only a technical element but also a cultural construct that can contribute to equity. The participatory and engagement mode further develops the argument that smartphones can provide greater access to storytelling and be used as a community engagement strategy.

Joanna Ong's *Reel Health* showcases the process of participation and engagement via smartphone filmmaking. The work by the New York collective Think/Feel is co-created by ten medical students in Tanzania. In the United States, there is one doctor for every 300 people; in Tanzania, there is only one for every 30,000. The Tanzanian medical students provide an insider perspective on the health situation from their point of view (POV). A similar co-created approach and production process is portrayed in Max Schleser's smartphone video *Massey Time Capsule*, which focuses on the time capsule's production process showcasing the container and how images are stored through a timeless computer algorithm, which will be opened in 2062. This online smartphone video is part of a digital strategy and demonstrates the potential to create attention online via video platforms and social media to drive engagement through storytelling through the process of a creative project.

Anne Massoni's *Yours and Mine 2* is using online interaction via picture messaging and the classical selfie as guiding principles for her work (Figure 2.29). Part diary/pen-pal

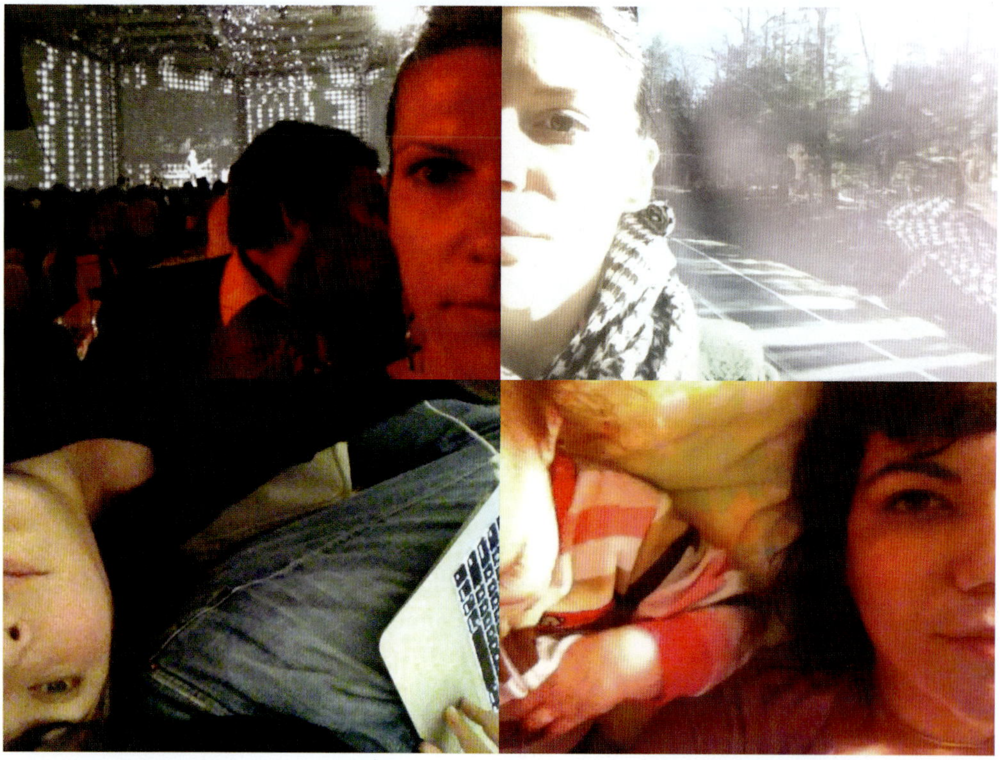

Figure 2.29 Still image: *Yours and Mine 2* (Massoni 2007).

exchange, part reflection on social media, *Yours and Mine 2* shares the lives of four women in 2006/2007 in the United States. Over the course of one year, more than 2,000 videos were combined by an algorithm into one work. In a similar participatory style *Here and There* is a collaboration between Morgan Royle in Manchester, UK, and Felipe Cardona in Bogota, Columbia, who use smartphones, tablets and apps to introduce mixed reality into the mobile-mentary production. Both filmmakers move the camera in a selfie-style 360-degree pan, which is intercut with changing locations in the background. In the video *#selfiesuit_documentation*, Paul Taylor combines smartphone footage shot by participants on four different walks through the streets of San Francisco and Oakland, California, United States, in 2016. Taylor wears a full mirror bodysuit, which positions the project at the intersection of performance, participatory art project and smartphone video. If passers-by showed interest, he encouraged them to participate by taking a selfie with him in the frame and using the suit's reflections to show the participants, who then posted their photos and videos via social media, thereby translating the project from the analogue world into the digital domain.

While the previous three projects describe new formations of creativity, related to the theoretical framework of participatory culture (such as Gauntlett 2011, Jenkins 2009 and 2019), *In Response We Closed Flinders* demonstrates the relevance of community commentary as social engagement beyond a mainstream media discourse. In 2016 rallies were held in the City of Melbourne, Australia, against the government's plans to close indigenous communities. Alex Dick stumbled across this rally on his daily commute, having seen a Facebook post about it on his smartphone. His mobile-mentary shares his experience of the events. While he did not see it as a political comment but rather a document of the day, his smartphone video shared his experience and engages his peers as observers. There were few comments of this event in the mainstream news, and social media provides an alternative space to reach audiences and make personal statements creating alternative discourses. Patrick Kelly's *Give Me Faces and Streets!* merges the online Google maps environment with a mixture of audio recordings using the Voice Memo app on an iPhone 6 in and around Melbourne, Australia, as well as small snippets of found audio (Kelly 2016). The flâneur-like drifts through online and offline spaces are a vital characteristic of the everyday application of mobile devices and smartphones, browsing either through the streets with GPS data or in the online world collecting memory artefacts and snapshots of locations. The participatory and engagement mode embraces the shoot and share approach, blurring the lines between producer and user. More significantly, it enables communities to create personal representations according to their perspectives.

2.6.3 Smartphone native mode

The smartphone native mode is a reaction towards the proliferation of apps for mobile devices, smartphones and connected pocket cameras with Android OS.[15] This mode is characterized by projects filmed and edited on smartphones (as opposed to laptop or desktop computers). There is a conceptual difference when editing the work on location and not in an editing suite. One is still immersed by the environment and when editing on the move. The notion of a Situationist's dérive (Debord 1956) can facilitate experimentation in filming and now also editing on location. Being less concerned with the revolutionary elements of the Situationists, the dérive is understood as a creative strategy which foregrounds unplanned encounters and journeys through a landscape and/or cityscape (online and offline) by intuition and emotion. In a similar way, as artists are taking the easel out of the studio to paint outdoors, 'en plein air', the native smartphone films discussed in this section explore a new subset of mobile art films produced on the go.

The smartphone native mode exemplifies this through mobile moments (Keep 2017) and ephemeral snapshots produced by Adrian Miles, David Scott Leibowitz, Max Schleser, Vanessa Vox and others. These creatives translated their backgrounds into the domain of mobile, pocket and smartphone filmmaking; Adrian Miles was one of the first video bloggers, David Leibowitz is author of *Mobile Digital Art* (Leibowitz 2013) and co-founder of the *Mobile Art Conference* in New York from 2010 to 2012 and Vanessa Vox is a visual artist who has been mainly working with mobile devices and apps since 2012. She uses the ability to intuitively transform the spontaneous shot from her iPhone to an icon- or dreamlike representation.

Day at Amagansett Beach is a one-minute trip to the beach, shot entirely on the iPhone using the app 'Infinite Eye' (Figure 2.30). During the early to mid-1970s, David Scott Leibowitz created many hours of abstract video art in a dark room by pointing a video camera at a monitor creating video feedback. This app creates a similar effect that you can

Figure 2.30 Still image: *Day at Amagansett Beach* (Leibowitz 2012).

take outside into the real world, to create video art using slices of reality combined with this abstract yet controllable visual art generator. Now, this artistic skill set that he acquired thirty-seven years ago as a video artist, manipulating video feedback, is available on Apple's App Store. The short film *David Hockney at the Met* by David Scott Leibowitz pays tribute to the most famous mobile digital artist David Hockney and his *Large Interior*, Los Angeles, 1988, hanging at the Metropolitan Museum of Art in New York City, United States, shot using an iPhone 5S and the app 'Time Piles'. As much as Leibowitz's mobile artwork, one can note that artists, filmmakers and makers are redefining the boundaries of both fine arts and popular culture alike. Another example of this wave of artistic and experimental mobile works that proliferated since the launch of the App Store and/or Google Play store is *Pacific Colours* (Figure 2.31).

With our smartphones, mobile devices and pocket cameras, we can all be virtual 'pop stars' finding our audience on social and networked media. *Pacific Colours* displays the vivid colour of the Pacific Island Tonga. Max Schleser edited the smartphone video on location, expressing his experience of being in a Polynesian environment through its unique colourscape. To capture the mobile filmmaker's experience and moment of its recording, the abstract kaleidoscope was chosen. *Pacific Colours* shares this moment with its audience. *Pacific Colours* points at not only the capacity for producing a video on the go but also the ability of music and audio which was produced entirely on mobile devices and smartphones using the 'iMaschine' music app.

Dreaming by Vanessa Vox is a short surreal, experimental movie made with iMovie, LumaFX, Artisto and Garage Band (Figure 2.32). She aims to blur the boundaries of our perception between dreamscape and experience. The dominant colours are blue and yellow, and camera movements follow her body movements from a third- and first-person perspective, which illustrates her feeling of being trapped in a mysterious image sequence.

Figure 2.31 Still image: *Pacific Colours* (Schleser 2017).

Figure 2.32 Still image: *Dreaming* (Vox 2016).

Next to surreal works, one can also note examples of more process-oriented experimentation.

5#CALLS by Giuliano Chiaradia was 100 per cent, entirely, scripted, recorded and edited on the director's own Nokia in 2011 (Figure 2.33). The scripts were written on the mobile phone and sent by SMS to the actresses Stresser Guta, Natalia Lage, Amanda Richter, Graziela Schmitt and Julianne Trevisol. Music was produced through the use of ring tones, and mobile phone wallpapers were used as graphic stills. The result is a collection of five videos that the filmmaker calls innovative experimentation in new media. The videos mirror video calls and the project uses mobile montages created via displaying images on multiple mobile phone screens. In 2013, Adrian Miles filmed *About 7 am: The First Quarter* using the no longer exiting app 'Vine'. The video hosting service for six-second smartphone videos was found in 2012, acquired by Twitter in the same year before the launch, before becoming discontinued in 2016. Adrian Miles filmed the same view, at about the same time and place every day, for a year using Vine. Another app that no longer exists is the

Figure 2.33 Film poster: *5#CALLS* (Chiaradia 2011).

collaborative video app 'Jump Cam'. This points at some of the difficulties when relying on cloud services, open frameworks and Application Programming Interfaces (APIs), which can change at any time. Another prominent example is the discontinuation of the YouTube Video editor. In the digital economy, it is not surprising to see new apps being launched, while others disappear.

Connectivity is not only a technical but also a social construct. *On the Move* (Felipe Cardona, Daniel Wagner, Laurent Antonczack, Max Schleser, Thom Cochrane) resulted out of a playful collaboration to make a music video. The co-creation process and the online collaboration demonstrate the network capacity of mobile devices and smartphones, allowing co-creators to work across multiple time zones and locations.

2.6.4 Conversational mode

The mobile phone is a ready 'off-the-shelf' networked camera. The fusion of communication and time-based media classifies the conversational mode. Odin argued that the cinema language has become part of our daily communication (Odin 2012). This mode goes beyond telephone voice or video-calling/conferencing and visual communication, understanding the interaction as a conceptual starting point to engage audiences in a conversation about a specific subject. Through the use of social media, these digital stories can be shared among peers, 'friends' and followers to be part of bigger dialogues. The conversational mode should also be described as a conceptual starting point to engage audiences and communities in a conversation. The bottom-up and horizontal approach to enable any maker to start a conversation about a subject that she or he is passionate about opens up the opportunities for more stories to be told in more diverse ways. The argument is not that an Instagram channel will create any social change but with a dedicated video content creator and community manager, social media can facilitate conversations, shape and engage communities and like-minded followers online.

How Was Your Birthday? presents us one of the most prominent crises in this decade. Through using a bird's-eye view, Mohammed Zaouche shows the extreme perspective of a young girl and how everyday life circumstances can change without one's own impact on the situation.

The work is a starting point for a number of conversations about the refugee crises and conflict war in the Middle East.

Two Stories by Zaher Omareen is a short documentary demonstrating the capacity for personal portraits and filmmaking. The filmmaker walks through a high street in the UK and films a long take, handheld in one shot pointing towards by-passers shoes and shoes in shop windows (Figure 2.34). The narrator, the filmmaker himself, uses an interview technique to tell two stories separated by ten years, addressing the issue of human rights violations under Assad's dictatorship in Syria. Omareen shares his childhood memory of being beaten up in the school court without the aggressors being reported due to their father's political status. The other story crystalizes around memory from a prison interrogation. All three elements are interconnected in the moment when the narrator tells us that the schoolyard aggressors and the kids on the playground had the same trainer shoes that he sees in the streets. This trauma still follows the filmmaker. His last words in the mobile-mentary are 'I wanna see human faces'.

Figure 2.34 Still image: *Two Stories* (Zaouche 2017).

Tofu Man by Andrew Robb is a portrait of Duc Nga, a former Vietnamese refugee who came to Australia, seeking a better life after being orphaned at fourteen years of age. Weaving Duc's story with his tofu-making business makes for a film of surprising poignancy. While Duc is making tofu, Andrew asks him about his past experiences and through Duc's tattoos, we learn about his past and his values: 'happiness is more important than money'. The conversation between filmmaker Andrew and Duc is present in the short documentary and more significantly demonstrates the potential to feature people from all walks of life, sharing their experiences and stories (Figure 2.35).

The conversational mode can also relate to the non-verbal mobile expression and/or performance, which is exemplified through the work of Camille Baker. Her *Ephemeral Mobile Media VJing* uses a live database approach which remixes body data from participants with wireless sensors to create abstract visual conversations with other mobile users and participants creating a collaborative, telematic collage of externalized body sensations. The video is presented as an abstract remix of the participatory-generated videos.

Mobile Narratives is told through the text-based interface and still images. Carey Scheer explores the invisible inner world of the phone user. It is a common sight to see people in public, sitting, standing or walking in silence with their eyes glued to their phone. This film taps into the thoughts and emotions running through the phone user's head that is anything but silent. This smartphone film shows the creation and then dismantling of a narrative that the main character created while communicating through her phone's interface. Another example of the blurring of online and offline experiences is Brice Veneziano's *Emergency Call – Appel d'urgence*, which intertwines video posted on social media and personal memories. His film draws upon child abduction through the perspective of a

Figure 2.35 Still image: *Tofu Man* (Robb 2014).

security officer at a Christmas market, who comes across the social media post of a father who is looking for his missing son. *Emergency Call – Appel d'urgence* exemplifies how the conversational mode goes beyond voice/visual communication drawing upon one's personal memories and interconnects these with our everyday social media engagement.

2.6.5 'Red carpet in your pocket' mode

'Red carpet in your pocket' mode engages with narrative forms of smartphone filmmaking. Films in this category deal with the development of linear storytelling and updating of classic narrative elements. Several filmmakers have used the phone as a storytelling device to let a story unfold or augment the various elements of storytelling such as suspense, transformation and resolution. In comparison to the other modes, the entertainment factor is its defining characteristic. This mode translates filmmaking as we all know it from Hollywood into an accessible endeavour, turning friends in crew and cast. The interview in Section 4.5 with Conrad Mess demonstrates that smartphone filmmaking can also provide an avenue to success.

Xavier Satorra's *Calling Soon* revolves around a real-life stalker game. Multiple victims receive a phone call threatening them not to hang up or otherwise their life would end soon. The short film is edited like a trailer and raises the suspense level through an anonymous caller, who acts as the antagonist. On receiving the call, the protagonist's emotions change during the phone call. Natalia's facial expressions and the short film non-diegetic music intensify the suspension. A close-up of her face intercuts with other mobile phone callers who are gamers in this worldwide conspiracy. A tear is rolling down Natalia's face to express her hopelessness, as we hear the instruction 'entertain the audience their votes can save you'. In a similar way, the feature-length work *God in My Pocket* by Arnault Labaronne uses the phone to create an eyewitness perspective for the audience. A mysterious man kidnaps the lead character, Caroline. Locked up in a dark apartment, Caroline uses her mobile phone to communicate with her kidnapper via video messages. This 2011 production uses the vertical video 1:1 format to create an authentic account. The director characterizes his story construct as 'terrible drama' and draws heavily on telephone conversations and Bluetooth messages between the kidnapper and locked away victim. Towards the end, one realizes that Caroline has been held captive by a male character who refers to himself as God and wanted to make a movie out of Caroline's lived out nightmare. In the final sequence, Caroline is freed by the director and kidnapper. This sequence is shot on a boat, and she throws the camera phone into the ocean, taking her revenge. The film ends with the underwater footage and the battery dying. The low-resolution aesthetic creates a connection to and presence for Caroline. The narrative and communication that one engages in are merged into one experience that feels intimate and immediate (Figure 2.36).

The Editor by Chris Nong is a meta-narrative construct. The story is intercut as a crime detective-like story with the life of an overworked movie editor. Chris Nong cuts between the parallel stories through the video editing interface. We drift between script, rough cut and the editor's life. An unpredicted twist occurs when the story shifts to a meta intervention,

Figure 2.36 Still image: *God in My Pocket* (Labaronne 2006).

Figure 2.37 Still image: *The Life and Death of the iPhone* (Trillo 2015).

with a director giving an editor feedback on the above story construct. This movie in a movie construct is similar to the social media representations we create with our smartphones on a daily basis.

The Life and Death of the iPhone by Paul Trillo follows the entire lifespan of a phone in POV from its creation in a factory into the hands of a New Yorker through its death and eventual rebirth (Figure 2.37). The short was both shot and edited on the iPhone using the Cameo – Video Editor and Movie Maker app, which Vimeo acquired in 2014. The short film takes the perspective of the iPhone's front camera. It starts with its assembly in China and Paul unpacking his new iPhone. The drama provides a snapshot of a New Yorker's professional life including a phone robbery and the classic phone drop into the toilet accident. The narrative brings the iPhone back to China, where it is fixed and sold again. Paul Trillo used the mobile characteristics of iPhone camera to film in several typical NYC locations such as public transport, museums, Grand Central Station, restaurants and bars, which provide a very authentic New York City experience (Figure 2.38).

In *Bubbles Don't Lie* Stepan Etrych portrays the engineer Cmiral (Figure 2.39). One morning a comic like bubble appears over his head (Figure 2.40). Inside this comic like speech bubble, the number six is visible. He goes on with his daily life, and news reports show that bubbles have appeared above everyone in the world, though no one can uncover the bubbles' meaning. The juxtaposition of various characters in a humorous way hints towards their meaning relating to the sexual encounters in the peoples' lives. While some stereotypes are confirmed, others give the characters' appearance a funny twist. The humorous work speaks to a serious subject, as personal information and data could get into the public domain through search engines, apps and social networks.

This short film also foreshadows the application of AR via the speech bubbles and highlights the sensitivity of personal information in public spaces while referencing the

Figure 2.38 Production still: *The Life and Death of the iPhone* (Trillo 2015).

Figure 2.39 Production still: *Bubbles Don't Lie* (Etrych 2015).

Figure 2.40 Still image: *Bubbles Don't Lie* (Etrych 2015).

smartphone's intimate characteristics. *Bubbles Don't Lie* as much as the other above-mentioned short films play with the audience's expectation of a narrative and gives these a twist and thus update classic narrative formulas. The 'Red carpet in your pocket' mode brings smartphone filmmaking to the masses and thus contributed to the mainstream recognition of mobile and smartphone filmmaking.

Notes

1. Blackmagic Design Pocket Cinema Camera, https://www.blackmagicdesign.com/au/products/blackmagicpocketcinemacamera
2. Arri ALEXA Mini, https://www.arri.com/en/camera-systems/cameras/alexa-mini.
3. Red Hydrogen camera phone, https://www.theverge.com/circuitbreaker/2017/8/2/16084142/red-hydrogen-one-holographic-phone-specs-photos
4. *N.I.* short film, https://www.nishortfilm.com/
5. ATOM (Australian Teachers of Media) education shop, http://bit.ly/MINAstreaming
6. Mobile Innovation Network and Association, https://www.mina.pro/
7. The FILMOBILE project was part of the Node London network which Graham and Cook discuss in the chapter 'Collaboration in Curating' (Graham and Cook 2010).
8. See Chapter 6, 'Creative Innovation'.
9. MINA panel at ACMI (Australian Centre for the Moving Image), https://www.acmi.net.au/events/international-mobile-innovation-screening-and-film-festival/.
10. Korsakow, http://korsakow.tv/formats/korsakow-film/.
11. For further information, please contact Max@mina.pro.
12. *#Filmbreaker* movement, https://www.facebook.com/filmbreaker/.
13. MINA Facebook group and page, https://www.facebook.com/mobileinnovationnetwork.

14. FilmFreeWay, https://www.filmfreeway.com.
15. Such as Nikon *COOLPIX S810c*, Samsung *Galaxy Camera 2* and *Galaxy Camera NX*, Polaroid SC1630 and Polaroid iM1836 or Yongnuo YN450.

References

Adweek. 2015. 'What the Rise of Native Video on Facebook & Twitter Means for Brands', *Adweek*. Available online: http://www.adweek.com/digital/what-the-rise-of-native-video-on-facebook-twitter-means-for-brands/.

Amerika. 2008. 'Immobilité – A Feature Length Mobile Phone Art Film', *MarkAmerika*. Available online: http://markamerika.com/news/immobilite.

Aston, Judith, Gaudenzi, Sandra and Rose, Mandy, eds. 2017. *I-Docs: The Evolving Practices of Interactive Documentary*. New York: Wallflower Press.

Astruc, Alexandre. 1948. 'The Birth of the New Avant-Garde: Le Camera-Stylo', in Peter Graham and Ginette Vincendeau (eds), *The French New Wave: Critical Landmarks*, 2009, 31–8. London: Palgrave Macmillan on behalf of the British Film Institute.

Atkinson, Sarah. 2017. 'Mobile Cinema', in Stephen Monteiro (ed.), *The Screen Media Reader: Culture, Theory, Practice*, 197–219. New York: Bloomsbury.

Baker, Camille, Schleser, Max and Molga, Kasia. 2009. 'Aesthetics of Mobile Media Art', *Journal of Media Practice* 10 (2&3): 101–22. https://doi.org/10.1386/jmpr.10.2-3.101_1.

Berger, Sally. 2010. 'Maya Deren's Legacy', in Carolina Butler and Alexandra Schwartz (eds), *Modern Women: Women Artists at the Museum of Modern Art*, 301–9. New York: The Museum of Modern Art.

Berkeley, Leo. 2018. 'Tram Travels: Smartphone Video Production and the Essay Film', in Marsha Berry and Max Schleser (eds), *Mobile Media Making in an Age of Smartphones*, 25–34. London and Basingstoke: Palgrave.

Berry, Marsha. 2017. 'Mobile Filmmaking', in Larissa Hjorth, Heather Horst, Anne Galloway and Genevieve Bell (eds), *The Routledge Companion to Digital Ethnography*, 308–17. New York: Routledge.

Corrigan, Timothy, ed. 1999. *Film and Literature: An Introduction and Reader*. Prentice Hall: Upper Saddle River.

Debord, Guy. 1956. 'Ideology Materialized', in Knabb Ken (trans.), *Society of the Spectacle*, 2014, 113–19. London: Rebel Press.

Deren, Maya. 1965. 'Amateur versus Professional', *Film Culture* 39 (Winter): 45–6.

Gauntlett, David. 2011. *Making Is Connecting – The Social Meaning of Creativity from DIY and Knitting to YouTube and Web 2.0*. Cambridge: Polity Press.

Graham, Beryl and Cook, Sarah. 2010. *Rethinking Curating: Art after New Media*. Cambridge, MA: MIT Press.

Jenkins, Henry, ed. 2009. *Confronting the Challenges of Participatory Culture: Media Education for the 21st Century*. Cambridge, MA: MIT Press.

Jenkins, Henry, ed. 2019. *Participatory Culture: Interviews*. New York: John Wiley & Sons.

Leibowitz, David. 2013. *Mobile Digital Art: Using the iPad and iPhone as Creative Tools*. London: Focal Press.

MacEntee, Katie, Burkholder, Casey and Schwab-Cartas, Joshua, eds. 2016. *What's a Cellphilm? Integrating Mobile Phone Technology into Participatory Visual Research and Activism*. Rotterdam: Sense Publishers.

Nichols, Bill. 2001 and 2010. *Introduction to Documentary Studies*. Bloomington: Indiana University Press.

Odin, Roger. 2012. 'Spectator, Film and the Mobile Phone', in Ian Christie (ed.), *Audiences – Defining and Researching Screen Entertainment Reception*. Amsterdam: Amsterdam University Press.

Rees, All. 2011. *A History of Experimental Film and Video*. London: Bloomsbury.

Schleser, Max. 2008. 'FILMOBILE', in Marina Vishmidt, Mary Anne Francis, and Jo Walsh (eds), *NODE.London Reader II*, 155–9. London: Mute Books.

Schleser, Max. 2011. *Mobile-Mentary: Mobile Documentaries in the Mediascape*. Saarbrücken: LAP Lambert Academic Publishing.

Schleser, Max. 2014a. 'Connecting through Mobile Autobiographies: Self-reflective Mobile Filmmaking, Self-representation and Selfies', in Marsha Berry and Max Schleser (eds), *Mobile Media Making in an Age of Smartphones*, 148–58. London: Palgrave Macmillan.

Schleser, Max. 2014b. 'A Decade of Mobile Moving-Image Practice', in Gerhard Goggin and Larissa Hjorth (eds), *The Routledge Companion to Mobile Media*, 157–71. Routledge: New York.

Schleser, Max. 2016. 'MINA – Mobile Innovation Network Australasia', *Media NZ Media Studies Journal of Aotearoa New Zealand Digital Communities*, 16 (1). http://dx.doi.org/10.11157/medianz-vol16iss1id200.

Vertov, Dziga. 1922. '"Kinoks: A Revolution" and "From the History of the Kinoks"', in Annette Michelson (trans.), *Kino-Eye: The Writings of Dziga Vertov*, 1984. Berkeley: University of California Press.

Wenger, Étienne. 1999. *Communities of Practice: Learning, Meaning, and Identity*. Cambridge: Cambridge University Press.

Films

#selfiesuit_documentation. 2016. Dir. Paul Taylor. United States: independent production.

5#CALLS. 2011. Dir. Giuliano Chiaradia. Brazil: Set Experimental.

About 7am: The First Quarter. 2013. Dir. Adrian Miles. Australia: independent production.

Absent VIII a homage to Maya Deren. 2013. Dir. Anders Weberg. Sweden: independent production.

App. 2013. Dir. Bobby Boermans. Belgium: 2CFilm and Just Film Distribution.

Bubbles Don't Lie. 2015. Dir. Stepan Etrych. Czech Republic: Aquarius Pictures.

By Any Old Light. 2014. Dir. Jean Bourbonnais. UK: independent production.

Calling Soon. 2015. Dir. Xavier Satorra. Spain: independent production.

Dark Waves. 2016. Dir. Sven Dreesbach. United States: independent production.

David Hockney at the Met. 2014. Dir. David Scott Leibowitz. United States: independent production.

Day at Amagansett Beach. 2012. Dir. David Scott Leibowitz. United States: independent production.

Dreaming. 2016. Dir. Vanessa Vox. France: independent production.
Emergency Call – Appel d'urgence. 2015. Dir. Brice Veneziano. France: independent production.
Ephemeral Mobile Media VJing. 2011. Dir. Camille Baker. UK: independent production.
Falling on Cement. 2016. Dir. Leo Berkley. Australia: independent production.
Film Socialisme. 2012. Dir. Jean-Luc Godard. France: Vega Film.
France/tour /détour/ deux enfants. 1977. Dir. Jean-Luc Godard. France: SonImage.
Give Me Faces and Streets!. 2016. Dir. Patrick Kelly. Australia: independent production.
God in My Pocket. 2006. Dir. Arnault Labaronne. France: independent production.
Haunting Melissa. 2013. Dir. Neal Edelstein. United States: Hooked Digital Media.
Here and There. 2014. Dir. Morgan Royle and Felipe Cardona. UK and Columbia: independent production.
How Was Your Birthday?. 2017. Dir. Mohammed Zaouche. Algeria: independent production.
Immobilité. 2011. Dir. Mark Amerika. UK: independent production.
In Response We Closed Flinders. 2015. Dir. Alex Dick. Australia: independent production.
Italian Love Meetings. 2006. Dir. Marcello Mencarini and Barbara Seghezzi. Italy: Emage.
Kolacze. 2013. Dir. Wilhelm Jerusalem. Poland: independent production.
La Ligne Brune. 2016. Dir. Rachid Djaïdani. France: independent production.
Le Monde vu par mes jouets. 2011. Dir. Alain Fleischer. France: independent production.
Massey Time Capsule. 2012. Dir. Max Schleser. New Zealand: Massey University.
Max with a Keitai. 2006. Dir. Max Schleser. Japan: independent production.
MBE MBHELE Uma umama engeneme ingane iyezwa egzzini – When a Mother Cries Her Children Feel It. 2019. Dir. Aryan Kaganof. South Africa: Africa Open Institute.
Meshes of the Afternoon. 1943. Dir. Maya Deren and Alexandr Hackenschmied. United States: independent production.
Mobile Narratives. 2015. Dir. Carey Scheer. Australia: independent production.
Mobilearte. 2016. Dir. Gerda Cammaer. Canada: independent production.
Moscow Diary. 2011. Dir. Adam Kossoff. Russia: independent production.
Night Fishing (Paranmanjang). 2011. Dir. Park Chan-wook and Park Chan-kyong. South Korea: Moho Films.
Nausea. 2006. Dir. Matthew Noel-Tod. UK: LUX and Film London.
N.I. 2019. Dir. Sallyanne Massimini. United States: Palo Pictures.
On the Move. Dir. Felipe Cardona, Daniel Wagner, Laurent Antonczack, Max.
Olive. 2013. Dir. Hooman Khalili and Pat Gilles. United States: CaveScribe.
Pacific Colours. 2017. Dir. Max Schleser. Tonga: independent production.
Rage. 2009. Dir. Sally Potter. UK: Adventure Pictures Ltd.
RIDES. 2007. Dir. Elan Lee, Sean Stewart and Jim Stewartson. United States: Fourth Wall Studios.
Reel Health. 2011. Dir. Joanna Ong. Tanzania: Tigo.
Saudi Runaway. 2020. Dir. Susanne Regina Meures. Switzerland and Germany: National Geographic.
Schleser, Thom Cochrane. 2013. Columbia and New Zealand: independent production.
Selfie. 2018. Dir. Agostino Ferrente. Italy: Magnéto Prod, ARTE, Casa delle Visioni.
Six fois deux. 1976. Dir. Jean-Luc Godard. France and Switzerland: SonImage.
SMS Sugar Man. 2006. Dir. Aryan Kaganof. South Africa: African Noise Foundation.
Sotchi 255. 2010. Dir. Jean-Claude Taki. France: Apatom.

Speech Marks. 2004. Dir. Steve Hawley. UK: independent production.
Tamarindo Costa Rica MM 22062012. 2012. Dir. Felipe Cardona. Columbia: independent production.
Tangerine. 2015. Dir. Sean Baker. United States: Magnolia Pictures.
The 57'. 2013. Dir. Leo Berkley. Australia: independent production.
The Silver Goat. 2012. Dir. Aaron Brooker. UK: Pinball London Ltd.
The Big City. 2018. Dir. Evan Luchkow. Canada: Aeon.
The Editor. 2011. Dir. Chris Nong. Russia: independent production.
The Life and Death of the iPhone. 2015. Dir. Paul Trilli. United States: Cameo.
The Q. 2015. Dir. Leo Berkley. Australia: independent production.
Tofu Man. 2014. Dir. Andrew Robb. Australia: independent production.
Two Stories. 2013. Dir. Zaher Omareen. UK: independent production.
Unsane. 2018. Dir. Steven Soderbergh. United States: Regency Enterprises and Extension 765. 20th Century Fox.
Us Stop. 2016. Dir. Julien Tatham. France: independent production.
Why Didn't Anybody Tell Me It Would Become This Bad in Afghanistan. 2007. Dir. Cyrus Frisch. Netherlands: Stichting Filmkracht.
Yours and Mine 2. 2007. Dir. Anne Massoni. United States: independent production.

3

Spotlight on mobile and smartphone film festivals

Chapter Outline

Film festivals are fundamental for a dynamic and independent film culture. Especially for emerging filmmakers, short film festivals are springboards and entry gateways to the industry. In a constantly changing funding landscape, short film festival directors, producers and organizers need to explore corporate sponsorship, art and cultural funding alike. Some countries like France (Wilson 2014) or South Korea (Wilson 2014) support film festivals to provide a forum for films in their respective languages and consequentially advocate regional or national culture and identity. Academic scholarship has also explored the socio-economic dimension of film festivals as part of urban regeneration and polishing the image of a particular city (Crespi-Valbona and Richards 2007), attracting tourism (Derrett 2000 and 2003), as well as a formation of sociocultural identity projects (Soyoung 1998). The smartphone film festivals are as diverse as short film festivals operating with a number of different funding and creative strategies. Smartphone film festivals provide opportunities for filmmakers to share their work with audiences and engage in a conversation with them. In smartphone film festivals, audiences include peers as much as emerging filmmakers or cinephiles. The open and engaging approach can be compared to short film festivals such as 48 hours film festivals and guerrilla filmmaking competitions (such as 48hourfilm.com or www.48hours.co.nz among several other festivals around the world). Similar to short film festivals like *Tropfest*, which labels itself as the 'world's largest short film festival', is that smartphone film festivals provide opportunities for filmmakers, creatives and storytellers to produce new work. In addition, short film festivals support local film culture and provide a meeting point for audiences and filmmakers beyond larger national or international festivals. Some of the well-established short film festivals also have film markets, such as Clermont-Ferrand International Short Film Festival. Since 1981 the short film festival organized an annual event and in 2016 launched the short film market, becoming now the second largest festival in France (after Cannes). Several major film festivals such as Sundance (Park City, United States), Cannes Film Festival (Cannes, France), Tribeca Film Festival (New York, United States), Raindance (London, United Kingdom), Venice Film Festival (Venice, Italy), SXSW Film Festival (Austin, United States), TIFF (Toronto, Canada) or Berlinale (Berlin, Germany) feature short film programmes alongside feature film screenings. A number of these well-known short film festivals featured a mobile or smartphone filmmaking competition, such as Australia's longest-running short film festival, St Kilda Film Festival, which included a Siemens MicroMovie Award in its 22nd edition in 2005. The same German manufacturing company sponsored the Siemens MicroMovie Award at the twentieth International Short Film Festival Berlin, which was organized by interfilm Berlin and featured Felipe Cardona's work, who is interviewed in the following chapter. The first film festivals that included mobile filmmaking were Big Digit (United States, 2003), Nokia Shorts (in collaboration with Raindance Film Festival, UK, 2003) and Tampere Film Festival (Finland 2003) (Schleser 2018, 114).

As mentioned earlier, the Vancouver-based film festival submission platform FilmFreeway[1] lists thirty-one smartphone film festivals at the time of writing this book. Twenty-one of these appeared during the last three years. In an academic context, one can point at Duarte and de Souza e Silva (2014), Simons (2011), Schleser (2013 and 2014) or Wilson (2014) who discuss mobile and smartphone film festivals. Duarte and de Souza e Silva review Brazil's ARTE.MOV and Mobilefest. The Mobilefest ran in São Paulo from 2006 to 2010 and featured some filmmakers and artists highlighted in this book. The Mobilefest also participated via live video link with the FIL*M*OBILE events in 2008. Writing in *Between iPhone and YouTube: Movies on the Move?*, Simons reviews the early mobile phone film festivals and discusses their position with digital media culture, situating these in a space in between video platforms such as YouTube and the film festival circuit.

In his words, 'the primary function of mobile phone film festivals appears to be to incorporate a new mode of filmmaking into that of traditional cinema' (Simons 2011, 106). He sees the move by the film festival circuit as an 'attempt to draw a boundary between the professionals of the moving image making business and the rising tide of DIY filmmaking culture epitomised by YouTube' (Simons 2011, 106). In the last ten years, YouTube not only developed its own enterprise ecology with conventions like VidCon, TwitchCon or Social in the City, among others; the smartphone film festival scene is now organized by smartphone filmmakers, entrepreneurs, directors and cinephiles for the smartphone filmmaking community. The festivals that Simons described had a special screening, category or side event for mobile filmmaking. Now the spotlight is on smartphone filmmaking, and as this book demonstrates, smartphone filmmaking established itself as filmmaking mode in the contemporary mediascape. Simons concludes by asking 'whether the gradual dissolution of the mobile phone film festival is a sign that this battle, too, has been lost' (2011, 106). At the time Simons was asking this question, most initiatives were mainly driven by industry professionals that were interested and invested in this emerging space to scout prospects from a film industry perspective. The interviews in this section demonstrate that now a third wave of smartphone film festivals continue to work of the early mobile film festivals and showcase a distinctive set of characteristics, influenced by the film industry but not dominated by the industry trends and directions. While one could use a rhetoric of inclusion rather than opposition, according to Simon's terminology in 2011 'the battle' might have been lost, but *Smartphone Filmmaking: Theory and Practice* argues that this 'fight' is continuing, and smartphone filmmaking is stronger than ever before. In 2014 Schleser focused on an analysis of early mobile phone films presented at festivals in Germany and in collaboration with Wilson and Keep (Schleser, Wilson and Keep 2014) discussed mobile filmmaking with a focus on Australasia, discussing the MINA 2011 and 2012 screening and symposia that all three participated in. Wilson's *Cell/ular Cinema* (2014) is an ethnographic study of international cell cinema film festivals (2010–13) and investigates the engagement of 'knowledge communities' and describes these festivals as 'culturalizing events'. *Smartphone Filmmaking: Theory and Practice* further develops and updates this scholarship and continues these discussions. For this book, four different smartphone film festivals were chosen to engage in the different environments within these are operating. Furthermore, it was key to choose festivals that Schleser attended, featured his smartphone films, such as SF3 and ASIFF or where he conducted workshops, for example, Mobile Film Festival. In the case of Mobile Motion (MoMo), Schleser is connected online via social media and Facebook groups such as Mobile Filmmakers to the community and online discussions. It is most interesting to note that smartphone filmmaking festival directors and curators are also filmmakers, such as Si Horrocks from MoMo in Switzerland, Angela Blake *SF3* in Australia or Benoît Labourdette in France whose interview is listed in the next section.

3.1 The global mobile and smartphone film festival scene

Mobile and smartphone film festivals operate in very different environments. The Pocket Film Festival in France was publicly funded by the City of Paris, Société Française du Radiotelephone (SFR), a French telecommunications company, and Centre National du Cinéma et de L'image Animée (CNC), an agency by the French Ministry of Culture. As a National Centre for Cinema and the Moving Image, CNC promotes the production and promotion of French cinema. In 2010 Schleser's mobile feature film *Max with a Keitai* screened at the Pocket Film Festival in the Forum des Images in Paris (Labourdette 2010). The Forum des Images in Paris is a cultural institution for all genres and forms of film, video and moving image. It was one of the leading smartphone film festivals with three days of programming in sold-out cinemas. The Forum des Images hosts five movie theatres with the largest one having a capacity for 444 guests. Not only did French institutions engage in the festival but also ARTE, a French-German TV network that focuses on the promotion of cultural programming, was involved in the festival activities. In the United States smartphone film festivals are organized as a private enterprise. The International

Mobile Film Festival is run by Susy Botello who is cinephile in the film festival scene. She runs a podcast, SBP Podcast – the voice of smartphone filmmaking,[2] a blog, a meetup and the festival. As not all festivals make it past the five-year mark, it is notable to recognize her engagement for the community. In LA the iPhone Film Festival is now slowly rebranding to MoziMotion. The iPhone Film Festival was co-founded and directed by Ruben Kazantsev from 2010 onwards as a hybrid between online film festival and smartphone filmmaking events, such as the San Francisco with MacWorld, Hilversum in the Netherlands Institute for Sound and Vision and Mobile Movie Days as part of the Dutch Media Week (Figure 3.1). He is also a visionary filmmaker who produced the international smartphone film *Departure*. The storyline of *Departure* includes location shoots in France, Belarus and the United States. As an independent filmmaker, his budgets did not allow for him to fly to all these locations with a crew. Ruben communicated via email and Facebook and had a smartphone filmmaking crew filming in three different countries. The footage was uploaded to the cloud via file-sharing sites, and the US team downloaded the footage for sound mixing. The French team did the fine cut and final edit, while special effects were done in Spain. While the team never met in person, he completed the whole project in 2014 and *Departure* screened at the Festival du Cannes in the following year.

Mobile Motion in Switzerland is funded through the crowdfunding website Kickstarter and now via Patreon.[3] The film festival organizers also run meetups and run an active blog. The ASIFF is the first smartphone film festival in Ghana and demonstrates the potential to engage communities through filmmaking in Africa. Independent producer Michael G. Osheku runs

Figure 3.1 Mobile Movie Days and Dutch Media Week 2019. Interview by Cesar Majorana (left), Ruben Kazantsev – MoziMotion (centre) and Ruud van Gessel – Mobile Movie Days (right) (van Gessel 2019).

ASIFF and this year also launched the African Short Film Market, the Nollywood model of independent producers in the world of smartphone filmmakers. In Australasia MINA positions itself as a networking project between industry, institutions and community groups. In Australia SF3, the SmartFone Flick Fest, is an annual meeting for smartphone filmmakers.

Characteristic of the global smartphone filmmaking scene is the peer-training, mentoring and showcase element that smartphone film festivals provide by the smartphone filmmaking community for the film industry. In the chapter 'A Decade of Mobile Moving-Image Practice', Schleser discussed the 'affinity spaces – highly generative environments from which new aesthetic experiments and innovations emerge' (2014, 166). Rather than only focusing on high impact productions and 'bumps on seats' measures for screenings and festivals, it is important to recognize the virtue of these public events for their open space and engagement of the community. Next to celebrating smartphone filmmaking, the knowledge exchange and a supportive environment are established in smaller events. As outlined in the interviews in this chapter, festivals are quite active on social media and provide the necessary support and recognition for emerging filmmakers. Considering the long tail (Anderson 2006), as discussed in the next section by Benoît Labourdette, means that showcasing work and producing smartphone film might not lead to a monetary gain immediately, but in the long run, these are milestones in the carrier of a filmmaker creating connections, exposure and opportunities for personal and professional development. Smartphone filmmaking embraces the change that the film industry is being critiqued for.

As you can see from the interviews, gender balance is not a question in the global mobile and smartphone film festival scene. Furthermore, the argument that smartphone filmmaking is the most accessible filmmaking form and camera and that smartphone filmmaking can provide means to democratize filmmaking is crystallizing in these interviews. Smartphone filmmaking as a form of twenty-first-century storytelling opens up the debate about how to bring more diverse stories to the big screen. In this context, one also needs to refer to the work at universities and specific film programmes, which can contribute to providing a change from within the academy. Karl Bardosh, Associate Professor at NYU University Tisch School of the Arts, teaches his Cell Phone Cinema Class since 2009. In 2007 he initiated the International Festival of Cell Phone Cinema in India and most recently at the Festival de Cannes Film Festival in 2019 established the Karl Bardosh Cell Phone Cinema Humanitarian Award.[4] This chapter will also make a contribution to the area of film festival research and provide a space for the work of film festival directors to be recognized.

3.2 International Mobile Film Festival (San Diego, United States)

Max: Thank you very much for being part of the book *Smartphone Filmmaking: Theory and Practice*. The International Mobile Film Festival has been around for quite some time. It's one of the first film festivals that appeared on the international scene.

Susy: Well, thank you for having me.

Max: Can you remember what made you start the International Mobile Film Festival?

Susy: I almost remember it like it was yesterday Max. The reason I got into smartphone film festivals was because I've always been fascinated with storytelling. The reason why I thought about mobile film festivals was because I wanted to get everyone to really try, to make films using a cell phone camera, as with all the other cameras that we have available. One day I realized, I should create a film festival because I was taking part in film festivals myself, just not with mobile phones. And I thought, to get people involved in making movies with cell phones they would need a challenge, and the best challenge is a competition, like a contest and a film festival is exactly what that is. I wanted to make that happen, I didn't want to do it online, which was really the only option.

In 2009 I was working on a film on location in Big Bear, California. There was a guy there who had a script, a screenplay in his back pocket, and he was part of the crew. One day I asked him, 'Is that yours, what is that? What do you have in your back pocket?' And he said, 'Well actually this is a screen play. I like to show it to everybody because it's my dream to make a movie and be the director of a movie. But I don't have the funding. I don't have the camera, I don't have the network.'

I came home and I thought, that's it. I'm just going to make it. I know the flip cams, if you remember those, I was working as an editor and in video production at that time and I knew that flip cams looked okay on a big screen. And it was that moment, when I said to myself, I am sure cell phones will come out looking great on the big screen. Then, I launched it, I launched International Mobile Film Festival. And, and that's how that started.

Max: From your own experience and having run the International Mobile Film Festival for a number of years, is there anything that you could see where you would say smartphone film festivals are different from standard industry film festivals?

Susy: Yes, most definitely. The main difference in a mobile film festival is that, and I see this as a benefit to filmmakers, you are inspiring people to make films who have never really made films before. If you're a professional and you have made a film with a smartphone, and you go to a mobile film festival, you are going to have people around you who are not professional filmmakers, who are really going to look up to you and they are going to look for ways that you can inspire them, and how your way of making traditional films will fit into the mobile filmmaking world when it comes to storytelling, screenplay, storyboarding and distribution among other things.

Smartphone filmmakers have a lot to share and learn from each other. And I think that is actually a benefit from the smartphone film festivals as opposed to the traditional film festivals. Everyone already knows the industry, so filmmakers are not learning as much from each other as in a mobile film festival. That's at least what I've witnessed.

Max: Are you staying connected with filmmakers who made a couple of films for your festival?

Susy: Yes. You and I know that there are some people in the mobile filmmaking community that decided to come out and give it a shot and they have made one smartphone film, and they have made up their minds after that they don't want to do it anymore, for whatever reason. And that is fine. And then there are the attendees. They go to the festival because they want to watch these movies. In a traditional film festival, you do not have people going to festivals who are just going there to watch the movies and say, 'Oh I want to make a film because I happen to have a Red Camera at home too.' But when attending a smartphone film festival, you realize 'all these films were shot with the same device that I have in my pocket. And I can actually talk to these filmmakers that are here at the festival.'

Max: That is really great to hear this from your perspective as an organizer who has seen the development and trends over the last years. If you consider what happened in the last five or ten years in the world of mobile and smartphone filmmaking, how do you see the future?

Susy: What I find exciting is that there are more and more people that are starting to give it a shot regardless of whether they are professionals, Hollywood directors or independent filmmakers or just someone who has never even thought about making films.

There really are no rules. I know especially for you because you love the experimental filmmaking part of this. And there are the Hollywood filmmakers who are only doing this as a hobby or maybe not a hobby but a challenge, a challenge to them. They have been making Hollywood films, they did not get to Hollywood by just making one film. They have been making films and they spend a lot of money on it and they are thinking, 'This might be a cool way to challenge myself to creating that film or sharing that story that I don't care to get funding for and it will be more my story than something with a big budget.'

What I see that is really exciting about that is that it is not going to stop someone who does not have anything, but a phone, from also making a movie. These filmmakers are also being accepted in film festivals that are not just mobile film festivals. That creates a really exciting opportunity for someone who has never made a film before to grab their phone and do it – even without using special gear or knowing anything technical about the filmmaking process. I think the value of the story is now finally becoming even more important than the optics of the cinema.

Max: As a smartphone film festival or mobile film festival organizer who also actively builds the smartphone filmmaking community, how do you engage with this community and how do you see this developing?

Susy: Well, you know, I love the filmmakers. As you know, people will submit a film and you may never have heard of this person before. And then you get a feel for how hard someone worked on a film. I can almost get a sense of the person that made the film, the person behind the film. It's not always just based on the story; it's based on everything about that film.

So my engagement with the community is, I look at the film, I see a great film, that's how I select it, and then I engage with the community and I find some really incredible backstories to each filmmaker, which make it really great for me to connect with each and every one of those people. I do not always get to meet the smartphone filmmakers in person. They do not always make it to the film festivals, but that does not mean that we do not connect and have a professional relationship with them. When it comes to the filmmaking community, there is also a bit of feeling responsible or obligated more than responsible to help inspire someone who has never made a film before who was having trouble and maybe not that much self-confidence into even making another film. I almost feel like the filmmakers are calling out and saying: 'Is it good enough?', 'Am I good enough?', 'if my film is good enough, then I am good enough.' And I will take the time, to give them some feedback. I will take the time to make sure to do that if we do not select that film, I do not want them to walk away thinking, 'I'm not good enough, so I give up.' I feel as a film festival organizer, director and founder you have a mission.

You do not want to knock them down on their first try. I also think it is important that when we make the selections that we are careful how we announce that and not say, 'Only the best film makers or only the best films made it' or things like that because that is saying the opposite. Whoever did not make it or could not make it for whatever reason, we all have internal reasons why we cannot accept all of the films. So, we have to be careful not to, to be sending the message that whoever did not make it is not good enough.

Max: This really highlights the community aspect and the inspiration that can arise from your festival. Are there any community champions that you work with? Any filmmaker whose work was featured at the international mobile film festival over a number of times in the festival?

Susy: I think the personal stories that people share, it takes a lot for someone to share a personal story, to share a story that affects them. We have the Community Stories programme, where you have to be a part of the community to share that story. Those films, those filmmakers to me in a way are heroes because it is not as fun to share a story that affects you personally and put it out there in a film and then go from one film festival to another. But it sure is something that has a really good outcome, I believe. As long as it's done right, then of course that is sending a positive message and that can inspire other people.

So to me, my heroes are always those filmmakers because I have an affinity or attraction to documentaries as well. It does not always have to be a documentary to send a powerful story. This question is really hard for me because I have been doing this for so long. There are so many people that come to mind that have shared stories like that. Actually, even you shared that story when you came here with *Frankenstorm*, about the people with the hurricane in New Jersey (Figure 3.2). To me those are hero stories. And so it's hard for me to just single one or two people out every year, every film festival.

There is always somebody that stands out in the submissions each year, but I would say there's people like Brian Hennings, from Perth, Australia, who came here the first

Figure 3.2 Still image: *Frankenstorm* (Schleser 2012).

time with a film *Express* that was very meaningful for me (Figure 3.3). Then he came again the second year with another meaningful film *Focus*. I think it was great that he came out the first time and he connected with everyone (Figures 3.4 and 3.5).

Susy: And there are two San Diego filmmakers. One is actually an actor, there's Anthony De La Cruz and his daughter, Miranda. He came to our film festival in 2016 with a music video and decided that he was definitely coming back in 2017 with another film. During the Q&A panel there was a question about mobile film making for youth and how I wanted to encourage more youth to make films (Figure 3.6).

Figure 3.3 Still image: *Express* (Hennings 2017).

Figure 3.4 Still image: *Focus* (Hennings 2018).

Figure 3.5 Brian Hennings interviewed by Susy Botello, IMFF 2017, film: *Express* (Botello 2017).

Figure 3.6 Miranda June Mullings, *Charlie*, IMFF 2017, with Father Anthony De La Cruz in 'Take Five', during Q & A Panel (Botello 2017).

One of his daughters, Miranda June Mullings, was sitting right behind me and I turned to her and I said, 'If you would like to make a movie and bring it next year and it makes it, and your dad is making a movie next year, then you could both compete with each other and you could receive the red carpet treatment just like him.' And I think she was like ten at the time. And to my surprise, she did. She made a stop motion video and she came back the next year with her father and submitted a film, *The Bullying Problem*.

We give everybody the same treatment . . . everybody is basically equal. And so, there she was with all these other filmmakers and she was a part of that just like any other filmmaker. And she opened up as a person and it made a difference in her school. They played her film and she was on the local news here because I brought her with me. And then, the following year after that, Anthony was not able to make a film, because he is an actor. But his daughter submitted another film *Charlie*.

Max: San Diego's International Mobile Film Festival is creating the next generation of smartphone filmmakers.

Susy: Also, Mickey Harrison, who is someone that I worked with many years ago, who was in the film industry for over thirty years. When I first told her about this, she laughed at me because she has many trophies. She has a room full of trophies from film festivals, including Cannes. And I said, 'Yeah, I have a film festival.' She says, 'Oh, that's nice.' And I said, 'Yeah, and everything is shot with a smartphone.' And she laughed and I said, 'Don't laugh Mickey, I'm serious.' And she then said, 'Oh, well how could that be?' And

Figure 3.7 Mickey Harrison with Tim Russ aka Tuvok from *Star Trek Voyager* during IMFF 2018 (Botello 2018).

the following year, she decided to work on a smartphone film and made the iPhone film *Dingleberry*. And she has made one every year, and the latest film was *Aunt Tillie's Kitchen*, which screened at IMFF 2019. She's eighty-four years old now, by the way (Figure 3.7).

Max: Are there any other filmmakers that inspire you? Whether they are in the world of smartphone filmmaking or beyond?

Susy: Yeah, there are. *Blue Moon* shot in New Zealand by Stef Harris. I interviewed actors Jed Brophy and Mark Hardlow for the SBP Podcast and then I really got to know what wonderful people they are. And what is inspiring about them is that they are borderline, I mean, they have worked on Hollywood films and they don't consider themselves as Hollywood actors, but they have had the privilege of being cast in Hollywood productions with *Lord of the Rings*, *The Hobbit*, *King Kong* and a lot of other Peter Jackson films.

And they have also done theatre. The really cool thing that inspires me about them is that they are now ambassadors of the International Mobile Film Festival in San Diego because they actually are very open to interacting and being part of the smartphone filmmaking community.

I do not think I have ever seen Steven Soderbergh join into conversations on Twitter and things like that with mobile and smartphone filmmakers and willing to connect at this level, like Jed and Mark have. I find their work really inspiring and I really look up to them for doing that.

Max: The International Mobile Film Festival has been around for quite some time now and it is very much dedicated to storytelling. And you have also mentioned the red carpet, which is always part of your festival (Figure 3.8).

Susy: The red carpet was a big deal for my film festival from day one. And I know you know that because you witnessed me stressing that we have to have the red carpet. The thing is, the part that was very important for me and continues to be is to give the mobile filmmaker the same – and this is from the very first film festival for me – the same respect as any other traditional filmmaker receives at any film festival. Now, we cannot do something

Figure 3.8 Red Carpet IMFF 2019 'Red Carpet Extravaganza!' (Botello 2019).

like Cannes where we have a mile-long red carpet going across the street type of a thing. Or have everybody riding in limousines but the feeling of 'I have accomplished something that has to be respected, they are rolling out the red carpet for me, I am getting in front of this audience and they are taking my photos and they are giving me the microphone.'

Those sorts of things I believe are very important for the filmmakers to receive regardless of whether they have made films with a mobile phone, smartphone or with whatever camera. I think that is part of the filmmaking process.

Max: We also need to talk about the exhibition or maybe even like distribution process in that respect. What you have highlighted with the red carpet is that the final points of a project are not only about the export from the editing suite or smartphone and you have uploaded and submitted the project to a film festival, but it is also about presenting the work in public. Or as you might say once you walked over the red carpet, that is when your film has become part of the festival and the filmmaker joined the community of filmmaker.

Susy: The red carpet signifies the high respect in this industry. The Academy Awards does this big huge red carpet thing before people go into the theatre. In film premiers, the big film premiers also have a red carpet, you know you are special and only the filmmakers are allowed on the red carpet. No one else is allowed on the red carpet. That is kind of the exclusivity of your accomplishment. Part of the award is the red carpet. But then of course you know me, I like to do things differently. I like to make things more exciting and to really give it more of a 'we are not just screening your films in a theatre' type of a thing, but you get to experience the red carpet, which makes it more festive and special.

So making the red carpet a bit of a show and an entertaining show, where people really can almost party on the red carpet. And we have done that with the inclusion of the *Star Wars Steampunk Universe* cosplayers who are pretty famous here in Southern California. And they are the only cosplay group for Star Wars that is approved by Lucas film (Figure 3.9).

Susy: I have a lot of respect and a sense of loyalty to the other unique film festivals. And that goes without saying, you and MINA, the iPhone Film Festival by Ruben Kazantsev. But what it also does is, it also does give me that special sort of fondness towards you and the other pioneers. And also, to the filmmakers who were making the films before everybody else jumped on this because we paved the road for everybody else, so everybody else has had a smoother ride. And it was a bumpy road when we got on it.

There was a lot of chatter on social media about the tools, that you must have this gear, you must have that. It must be, you must know all this, and you must know all that. And I think it comes from a lot of the groups, in the conversations they are having with people who have experience in this film industry. It gets a little bit overwhelming and intimidating for someone who has got a dream, has a smartphone, and has heard of this and wants to go out and make a movie or who has tried it and made one movie, it wasn't so great, but they are thinking about making another one.

But they are hearing this chatter and it sort of holds them back. 'Well maybe I shouldn't, maybe my film will never be good enough because I don't think I can learn about all these terms, these filmmaking terms, all the technical aspects. I don't know that I can afford a Gimbal or anamorphic lens, or I do not know if I can.' You know what I'm saying? I think we need to simplify it and remember that the reason you and I and others created these film festivals was to give an opportunity to everybody that had a phone to realize their

Figure 3.9 *Star Wars* – Steampunk Universe, Red Carpet IMFF 2019 with founder Susy Botello (Botello 2019).

dream and get their foot in the door through smartphone filmmaking. But if we are now going to turn it into what the independent film and video production industry has done and complicate it, what that is saying is that people who have a phone cannot make films with their smartphones, because they feel like they cannot compete with all these people who are now a part of this that are way superior. And that sort of defeats the purpose of what we are doing, and I think that ends up hurting the community in the long run.

Max: Accessibility is one of the key elements for smartphone filmmaking. Your emphasizing this point is a fantastic way to round up this interview.

Susy: Thank you.

3.3 MoMo – Mobile Motion (Zürich, Switzerland)

Max: How did you get into organizing smartphone film festivals?

Si: I met Andrea Holle through a feature film I was self-distributing *Third Contact*. The film was shot on a consumer camcorder and I was using crowdfunding to show it in cinemas. One of the best screenings we had was organized by Andrea in Zürich, Switzerland. Andrea told me about her idea for a smartphone film festival. I agreed to help out with the first event and then never left.

Max: Do you think smartphone film festivals are different from film industry festivals?

Si: Well, there are many, many film festivals in the world and they are growing in numbers all the time. Apart from the big name festivals, there are festivals which are simply a city or town's main film attraction of the year. There are also specialist festivals for all kinds of genres of film and so on.

I don't see smartphone film festivals as inherently any different to other festivals. They simply focus on the camera used, which is quite unusual. Of course, it depends on the festival and I only know MoMo (Figure 3.10).

At MoMo, we try to work within our limits and best as we can. We have almost no funding, apart from our Patreon support. We are champions of resourceful filmmakers who work like us – with minimal funding.

We are more like a family than an industry event. I have only been to Berlinale once, where I also had a look around the European Film Market. It's a lot less personal and you can get lost in the crowd. I am sure it's even worse in Cannes.

But at MoMo, nobody will get lost. We try to give all our filmmakers and audience our equal attention. We are not here to provide a film marketplace. We are more like a nature reserve, protecting and nurturing talented filmmakers, providing a springboard into whatever future they find for themselves (Figure 3.11).

Max: Considering what happened in the last five or ten years what developments are you excited about?

Si: At MoMo we are seeing more and more serious filmmakers using their smartphones to make films – short films and feature films. In the future, I expect the novelty will wear off. Shooting a film on your smartphone will be no more remarkable than using a DSLR or an ARRI ALEXA.

Figure 3.10 Mobile Motion 2018 (Mobile Motion FF 2018).

Figure 3.11 Mobile Motion's audience 2018 (Mobile Motion FF 2018).

At that point, we will know if this medium is here to stay. Although we have seen Soderbergh make two iPhone feature films (*Unsane* [2018] and *High Flying Bird* [2019]), I wonder if he will make any more using smartphones. Or was he just jumping on the extra marketing value of using a smartphone?

Smartphone camera sensors are going to get better, with Samsung[5] just unveiling their 108 megapixel camera which can shoot up to 6K video at 30 fps (Samsung online 2019). So it seems as if the cameras will keep improving in quality.

Personally, I am only excited in the stories that will get told. And the technology should not make too much difference there. But the more people that choose a smartphone camera to shoot, the more great movies we will be able to screen.

Max: How do you engage with the community of smartphone filmmaking and how do you see this developing? (Figure 3.12)

Si: We have always been very active on social media. And in the last year we have been running a blog too, which attracts a lot of traffic to the festival. We will continue to expand, but there's a limit to what two people can do without proper funding.

Max: Can you share any best practice examples for organizing smartphone film festivals? Is there anything that you consider that makes MoMo original

(i.e. certain genres that are screening or programming additional events)?

Si: We try to keep it simple and focus on the filmmaker and audience experience. We put a lot of effort into communicating with our selected filmmakers, which is one of the reasons we are constantly ranked in the top 100 of festivals on FilmFreeway[6] (Figure 3.13).

Max: Any projects or people/practitioners/smartphone filmmakers that inspire you?

Si: I am inspired by all filmmakers throughout history, using all kinds of cameras. In terms of smartphone filmmakers, take a look at our previous winners, they are my inspiration.

Figure 3.12 Daphna Awadish at Mobile Motion in 2019 (Mobile Motion FF 2019).

Figure 3.13 Mobile Motion at UTO-Kino (UTO-cinema) in 2019 (Mobile Motion FF 2019).

Max: Are there some films that have inspired you?

Si: Thousands!

Max: Of course every project is a bit different because you work with different people, you work on different locations, you work in different environments, but is there something that you can think about your production process, in terms of when you get together with a crew, when you go on set, is there something that you have a feeling that you can see in all of your films being realized?

Si: I have always worked with a minimal crew. I have written, directed and shot most of my films so far. I try to improvise as we are filming, real guerrilla style filmmaking. I tend to let the actors do their thing and I work around them, whereas most films are made in the opposite way – the actors must 'obey' the camera (Figure 3.14).

Figure 3.14 Production still: *Silent Eye* (Horrocks 2018–20).

Figure 3.15 Production still: *Silent Eye* (Horrocks 2018–20).

I'm not someone who spends too much time polishing and perfecting every shot. I am like an anti-Kubrick. I prefer to be fluent, flexible, more like shooting a documentary. What one calls cinéma vérité, I guess. So smartphone cameras are perfect for this style of filmmaking (Figure 3.15).

Max: Do you think you can do some things with a smartphone that you might not be able to do with a different camera for instance? Have you come across some things where actually the limitation was an advantage for you?

Si: Of course. Every camera has a limitation. Every budget involves compromise. Are you compromising your creativity to suit the funders of your film? Or are you compromising the kit and crew list to maintain your creative vision. It's always a payoff between one and the other.

As I say, smartphones are flexible, allow great spontaneity, fluency, shooting on the fly, anonymity and so on. When they do these YouTube camera tests between a RED and an iPhone, they always focus on where the smartphone is weakest – that is: the quality of the image. Let's see a comparison when trying to shoot on the fly.

I'd like to see an ARRI ALEXA and crew jump on a train and film a scene without anyone noticing. Or in a cafe or on the street. You would soon draw attention and get moved on.

Max: Do you have a preferred smartphone set-up in terms of gear? Can you tell us maybe about which are your favourite tools? And what excites you about this?

Si: Sometimes I use nothing but the smartphone. Other times I use a three-axis gimbal. I rarely use a tripod or cage/grip, but have done in the last episode of *Silent Eye*. I did one episode with a Moondog Labs anamorphic conversion lens, too. I always use the FiLMiC Pro camera app.

Max: What software do you use for the post-production?

Si: Adobe Creative Cloud. Premiere, Photoshop, Audition, After Effects.

Max: Do you think you'll always stay with smartphones, pocket cameras, iPhone? Or you think you will shift at one point to a different camera?

Si: I'll just use whatever camera suits the project.

Max: If you think back about what happened since 2010 in the smartphone filmmaking can you look into the future and can you imagine what could happen?

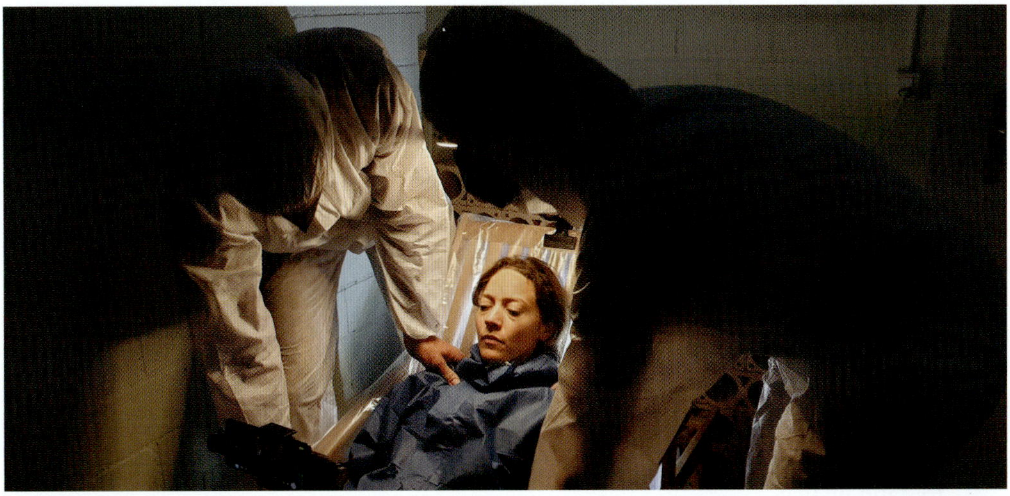

Figure 3.16 Production still: *Silent Eye* (Horrocks 2018–20).

Si: I'm sure more people will use smartphones to shoot films. The quality will improve. The line between DSLR, mirrorless and smartphone cameras will blur. People will forget there was ever anything unusual about filming with a smartphone.

Max: Is there anything that you would like to mention in relation to your work or smartphone filmmaking?

Be sure to check out our *Silent Eye* series which is currently free to view for Amazon Prime members (Figure 3.16).[7]

3.4 SF3, SmartFone Flick Fest (Sydney, Australia)

Max: Angela, thank you for joining and being part of the book *Smartphone Filmmaking: Theory and Practice*. It's really great to have you and SF3 a major international festival in the smartphone filmmaking scene being features in the book. SF3, the SmartFone Flick Fest, programme is expanding and now also smartphone feature films are screened and a dedicated kids filmmaking programme was initiated.

The first question is really sort of to hear a bit more about your story. How did you get into organizing smartphone film festivals?

Angela: It is not something I ever thought I would do, I guess it is really a fairly new industry. I come from a background of performing, I have worked my whole career as a performer. First, I was on stage as a singer, and a dancer, and then as I got older, I transferred over to acting and directing. I had just been living in LA for two years, and I came back to Sydney. And I directed a short play in the Short + Sweet Theatre Festival here in Sydney, and I auditioned an actress called Ali Crew, who happens to now be my co-founder of SF3, and so I chose her for my play (Figure 3.17).

We kept winning different rounds for the festival, we got all the way through to the big Gala Finals at the end, and so we spent about three months travelling through lots of rounds of the festival, and we were really kind of inspired by that festival, in that it gives anyone who has an idea for a play, or who wants to be on stage, or who wants to direct, it gives anyone a chance to have a go at that. And we just thought, 'Why is there nothing like this for film, especially in Australia?' And we kept thinking, and then I said to Ali, 'Oh, you know, when I was in LA, people had just started making films on phones, so it's something very, very new.' We said, 'let's combine those ideas', and then we thought we'll put on a Film Festival and give anyone with a smartphone the chance to make a film or be involved in filmmaking.

That is how it came about, we did not really know much about it or anything about it actually. We started researching and then we did, we found MINA, we found Susie in San Diego and Jason Van Genderen. I think you three were kind of the only. . . . You were the World leaders there. You were the only three kind of happening, and so we thought we would come in and take that space in Australia . . . . At the time you were in New Zealand, and we are Sydney-based, and we wanted to give everyone that chance to have a go, and so our motto is filmmaking that's accessible and affordable, and we still strive to stand by that.

Figure 3.17 Angela and Ali welcoming speech at SF3 2019 in the Sydney Opera House (SF3 2019).

Max: Leading on from your motto, do you think that smartphone film festivals are different from film industry festivals?

Angela: Yes and no. No, in that we have live screenings, handing out prizes, mentorships and provide networking opportunities. In that respect everything that we do including the laurels is like a traditional film festival, except films are shot on a smartphone (Figure 3.18).

I think we are different in that, because we are kind of trailblazing the smartphone film movement; I have not felt it anywhere else, there is a real big sense of community, and I think it's extremely strong in the smartphone filmmaking world. I find most filmmakers, and for most of us festivals aren't competitive with each other. I feel like we are all helping each other, and we are all excited to work in the space. Hence, you are putting together this book, and including everybody, and honouring everybody, and I feel like that is a perfect showcase of what this space is and the film festivals are. Yes, we are traditional film festivals, we're the same, but no, I feel like we are a bit more exciting, we are a bit more community driven, and I like to think we are a little bit cooler.

Max: I think there are some key characteristics in the world of smartphone filmmaking; you've mentioned accessibility; how about open forum and innovation? And at SF3 there were a number of feature smartphone films submitted to the festival?

Angela: Yes about eleven. You can just pick up a smartphone and start shooting stuff and making content rather than being a year in pre-production and all the heaviness that goes with that. It's a lot more vibrant.

Max: And considering what happened in the last five to ten years, what developments are you most excited about, if you look what could happen in the next few years.

Figure 3.18 Ms Jiyoung Park, WIFT (Women in Film & Television) International Winner (SF3 2019) (https://wiftaustralia.org.au/).

Angela: In SF3, we're really excited about the feature films that are being made. I told you, we had eleven entered.

It was our first year that we put the call for smartphone feature films out. For me this is really, really exciting that people are moving beyond the short form and actually making features on their phone. We screened *Blue Moon* by Stefen Harris. I was flying back from the Cook Islands, on Air New Zealand, a couple months ago, and I just happened to see that on the in-flight entertainment system. Now smartphone films are getting in flight, that is exciting. And nowhere was it written that it was shot on an iPhone, so it was not even, it was not even a thing anymore. I find that quite exciting as well. The films are so great and looking so great that you don't even need to tell people this is shot on a phone, and then when they find out they are amazed (Figure 3.19).

The VR and 360° is an exciting space. It is still very unrealized, and I am always excited by the films we get submitted every year. It's growing year by year and there is a lot of scope there. I do not know exactly where that's headed, but it is exciting to watch and be a part of. For me, we are really going back to that accessibility and affordability, we are really excited.

We are pushing really hard to grow our kids sections, that's for filmmakers sixteen and under. We're really passionate about master classes and encouraging young filmmakers, the next Phillip Noyce's and the next Jane Campion's, to pick up their phones and start making films (Figure 3.20). And then, they'll have this body of work, they will want to go to film school, they can go to film school and maybe they will just start making their own features, who knows. But, that to us is very exciting, and every year where we're really growing the amount of kids entries we get, and especially local kids entries from Australia, which we're passionate about, the Ozzie film scene. That for us is quite exciting (Figure 3.21).

Figure 3.19 Production still: *Blue Moon* (Harris 2018).

Max: There is a colleague in the Adobe Education Leader community, Joel Aarons, who is teaching filmmaking in a primary school. I think that is quite amazing. It is not only in high school that students are exposed to twenty-first-century creativity but in primary school taught about visual storytelling.

Angela: That is right. Imagine the confidence that you have, you're making films since you're in primary school, like you're so confident in making films, it's exciting for them.

Max: Next to the kids section, how do you engage with the community of smartphone filmmakers, and how do you see this developing?

Angela: We are really big on social media, Ali and I are always available to chat. It is a community, if anyone ever writes us email, Facebook, Insta, Twitter, we will always reply. A lot of people send us their films for feedback, or if we see filmmakers that might need motivation, we're big on motivating filmmakers, because we just want to empower everyone. Ali is a journalist and actress, both of us come from artistic backgrounds, we know what it's like to see a project develop from ideation to premiere. We are always open for anyone to write to us at any time and we will reply.

We organize masterclasses and we get hired by other people to conduct smartphone filmmaking classes. We have a partnership with NIDA (National Institute of Dramatic Arts) here in Sydney, we do workshops with them, we also provide teacher training for them. We do a lot of work in museums working with curators and with high schools. So, that's building the community (Figure 3.22).

As part of the festival this year, we had two networking events for filmmakers and sponsors to all meet each other. Film teams were formed on networking events and then completed projects came back in the following year and made it into the official selection. We also run master classes as part of our festival, and then for our prizes we source mentorships with key people. We want to build this in hooking up our contacts in the film world to our smartphone filmmakers and trying to bridge the gap between screening in film festivals and working full time professionally. That is something we are focused on with mentorships and introductions.

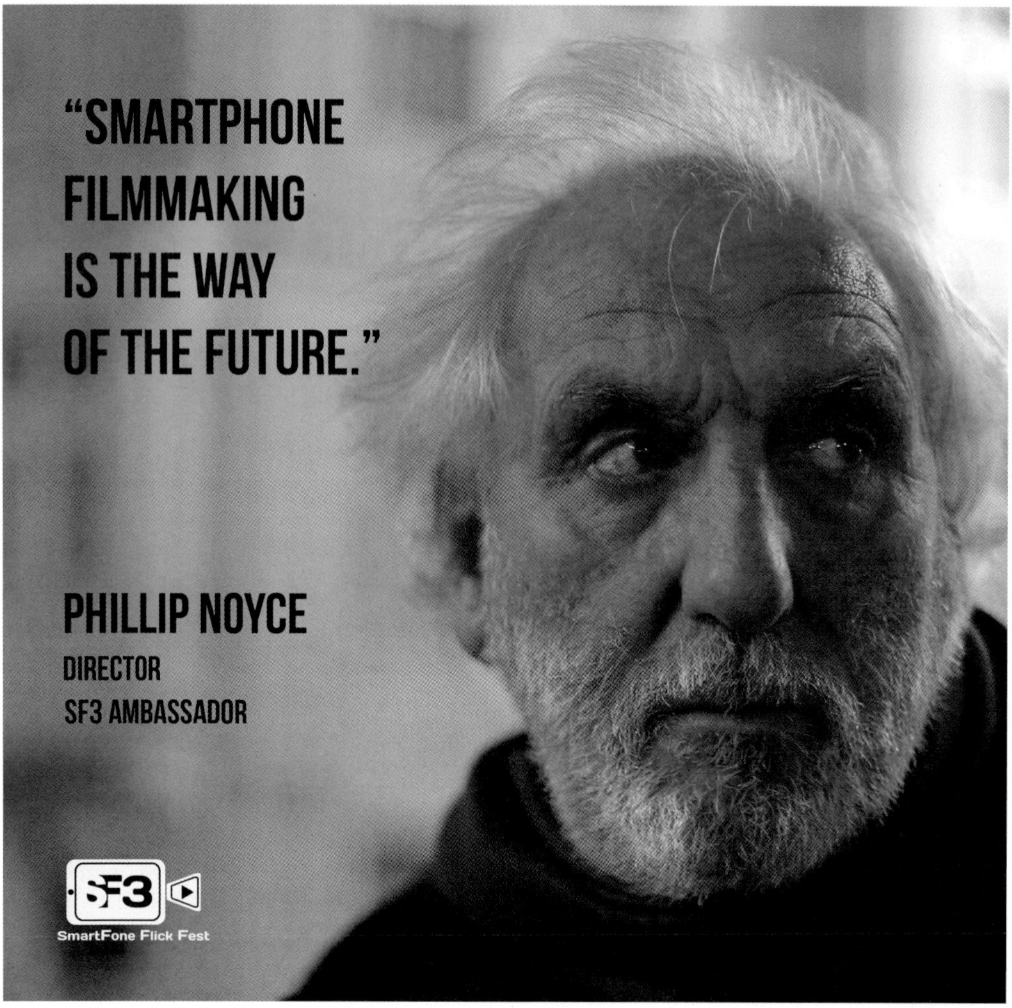

Figure 3.20 Phillip Noyce, SF3 ambassador. 'Smartphone filmmaking is the way of the future' (SF3 2019).

Max: How do you engage with your SF3 smartphone filmmaking community and are there any community champions that you can think about?

Angela: We definitely have very strong relationships with quite a lot of filmmakers and there are a few filmmakers in New South Wales, which for three or five years in our first year produced smartphone films for SF3, and they have been making films for us ever since.

They don't always make the finals though, so we do not show any favouritism to these filmmakers, but we have very close relationships with them. Filmmakers like Adrian Jeffs and Jason Van Genderen, who is one of our major sponsors, need a mention. Narelle Nash being a Sydney-based one, Brian Hennings and the list goes on. Christopher Stollery, who won our festival in 2016, is now one of our ambassadors. We kind of like to keep working with these filmmakers who are really passionate about smartphone filmmaking and keep encouraging them and finding opportunities. Adrian Jeffs made his first film *Good Day to Die* for SF3 in 2016 and now he's working professionally and doing amazing films (Figure 3.23). Another award-winning filmmaker is Ren Thackham who

Figure 3.21 SF3 Kids Best Director and Best Screenplay Award Winner Kate Snashell (SF3 2019).

won most of our major awards in SF3 2017 and best screenplay in 2018. She has now moved on to feature films after perfecting her craft in shorts, some shot on a smartphone.

Max: The next question I have, are there any projects or practitioners that inspire you? From the world of smartphone filmmaking or beyond?

Angela: First we are inspired by all the other smartphone film festivals because it can be a lonely thing running a film festival, you kind of spend most of your time emailing or hustling and doing things all by yourself, for very little time with other people and especially with other people who run film festivals. I always love catching up with other festival directors and just talking about the business, because it's something you don't get to do very often. Watching all the other festivals is very inspiring. Your festival is so international, screenings in China and New Zealand, that's very exciting. You and Susie were right there at the beginning, so I find that really inspiring. You got MoMo and Dublin and Toronto doing amazing things in all their respective countries.

We are also really inspired by what everyone else in the smartphone filmmaking world is doing. Jason Van Genderen, as I mentioned, is a huge inspiration for us here at SF3. He is one of my most favourited filmmakers and he just happens to shoot on smartphones. He is very talented and just such a lovely guy and willing to impart his masses of knowledge to anyone; he is very generous. He is a big inspiration for us here at SF3 and for me personally. Who else inspires me? It is a great question. I get really inspired by our filmmakers. Everyone has a story and I form very strong relationships with a lot of our filmmakers. And when you hear their stories it is always very interesting and their reasons for making films and things they have had to overcome. I get a lot of inspiration from just talking with a lot of smartphone filmmakers.

Figure 3.22 High school smartphone filmmaking workshop production still (SF3 2019).

Figure 3.23 Still image: *Good Day to Die* (Jeffs 2016).

Max: As a festival SF3 is now going for five years and the festival is going really strong. You screened in the Sydney Opera House last year and this year SF3 is at major picture houses in Sydney. As you have got a good overview of what happens at other smartphone film festivals, what sort of quality or characteristic do you think is very SF3 like.

Angela: That's a good question.

Max: Having been at SF3 and having had the great joy to participate in two screenings myself, is there something that you identify as SF3 selection criteria. Every festival has a different jury and every jury has a specific dynamic. And every festival has a slightly different curatorial process. Is there something that you identify as a SF3 'speciality'?

Angela: SF3, we are an international festival with an Aussie flavour is kind of how we think of ourselves and we are really focused on story and great storytelling. Ali works as a producer and a journalist at ABC in her day job. I have an undergrad degree in journalism and I'm currently studying for my master's in creative writing and screenwriting. So you have got two women running this festival who are extremely focused in their daily lives. We definitely look, number one, for a great story. That's number one for us. I know that doesn't set us apart, there are quite a lot of other festivals that focus on story, but it's very, very strong for us is story.

Max: This is a great insight. The festival leadership team is composed of two women. Considering that the Australian film industry as much as the film industry internationally isn't always the most gender-balanced industries. One could point this as a characteristic in the smartphone filmmaking world. SF3 demonstrates that smartphone filmmaking can be more inclusive than the film industry, because it doesn't really matter that much where you are from and what your background is. But if the smartphone films are outstanding, they should be celebrated at festivals.

Angela: We spend a lot of time on the curation . . . . Because you have to programme the films that you want, but you have to programme a screening for an audience. We do spend a lot of time programming, curating a programme for an audience as well. There are a lot of things and factors that go into curating a festival. We do try to pick the best films and we do try to pick the best experience for the audience as well.

Max: I had the feeling that SF3 celebrates great stories. You have been doing very well with the audiences as you always have a sell-out screening year after year. It is fantastic to see SF3 positioning itself in the Sydney film festival landscape. SF3 is recognized not only in Australia but also internationally.

Max: Let me say thank you again for a really great contribution. Smartphone film festivals are quite important to bring people together. Organizing a festival is a major event and arts endeavour, and the contribution you make to smartphone filmmaking community is outstanding. I think it's an important space, not only to watch a film but also to get together and discuss smartphone films.

Angela: Thank you Max.

3.5 African Smartphone International Film Festival (Lagos, Nigeria)

Max: How did you get into organizing smartphone film festivals?

Michael: I am a filmmaker who is passionate about telling compelling African stories to an international audience. I see a lot of aspiring and professional filmmakers always struggling to make a film, and the challenge is the budget. In Africa, most filmmakers cannot access film grants or funds to make a film. So now the question is, how do you become a filmmaker or a better one when you cannot make a film, due to no resources to do it? I saw a problem that needs a solution and I thought to myself that I can make a change in Africa by changing the narrative that great film must come with big budget, you can make a film with limited resources and it can start with a device that is readily available to you, the smartphone. In 2017, I founded the first smartphone film festival in Africa, called the African International Smartphone Film Festival, a platform that showcases and celebrates amateur, budding and professional filmmakers to share their project shot with smartphone to the world, irrespective of budget, quality or industry connection. Our first outing was very successful and we got submission from over sixty countries and we were able to screen over 200 short films made entirely on mobile device that is the smartphone, Tablet and PC. In Africa, we introduced and featured the works of thirty first-time filmmakers who were inspired by the festival to make their first film, and the numbers keep increasing each year of the festival. Today, we are listed among the world's best smartphone festival and one of the best-reviewed film festival on FilmFreeway,[8] a platform where filmmakers submit their project to film festivals and we in partnership with different brands and film festivals worldwide (Figure 3.24).

Max: Do you think smartphone film festivals are different from film industry festivals?

Michael: Film festivals are platforms for filmmakers and distributors, which serve many purposes from being a marketplace to promoting emerging talents and social causes. Smartphone film festivals are organized in the same standard like every other film industry festivals in the world, and the difference is that they showcase films shot entirely

Figure 3.24 Michael Osheku, founder and festival director of African Smartphone International Film Festival (right), and Oladipo O' Fresh, filmmaker and workshop facilitator (left) during ASIFF 2019 (ASIFF 2019).

on mobile devices and most smartphone film festivals that I know screen only short films not more than 15 minutes and run from one to three days. And there are smartphone festivals like ours that screen 360° virtual reality films. Filmmakers who submit their projects to the festival are still experimenting with smartphone and they prefer to make shorts. Smartphone film festivals are growing rapidly and new festivals are springing up yearly; we cannot say just yet that we have a bigger mobile film festival that can be compared to any of the major film industry festivals. But one thing that mobile film festivals has been able to achieve over the years is disrupting the film industry and film festivals' space by providing a unique platform for emerging smartphone filmmakers – this is ground-breaking and it's what brings us to the future.

Max: Considering what happened in the last five or ten years what developments are you excited about? (Figure 3.25)

Michael: In the last ten years smartphone filmmaking is not popular or close to non-existence in Africa.

I am excited about the awareness among creative talents using this simple tool, the smartphone, which opens new way of telling stories with limitless possibilities, and now we have professional smartphone filmmakers creating amazing content with just a pocket device. Is that not awesome? (Figure 3.26)

I am also excited about the technologies available for smartphone filmmakers to exploit, which help them in making beautiful images, better audio/sound, efficient and flexible software for editing, colour grading and special effects. Online content creators now opt in for mobile filmmaking.

Recently some young Nigerian filmmakers went viral for shooting sci-fi movie with a smartphone, and they made headlines on popular international television stations, magazines and blogs.[9] If we have not made this statement in Africa, these young talented kids will still be aspiring to become filmmakers. Smartphone filmmaking has been able to take aspiring filmmakers from the point of aspiration to the point of actualization. And

Figure 3.25 Live recording of festival activities at ASIFF 2017 (ASIFF 2017).

Figure 3.26 Raymond Yusuf, Critics Company (extreme left), Cletus Clement, director of *Backwaters*, Oladipo O' Fresh, Michael Osheku, Judith Audu, member of Screening Committee, Bolatito Sowunmi, Ivan Imoka and Godwin Josiah, Critics Company, during award presentation at ASIFF 2019 (ASIFF 2019).

now, many social media content creators and influencers are making use of their smartphone to make content. In years to come, DSLR cameras will no longer be in the market as people will not need them.

Max: How do you engage with the community of smartphone filmmaking and how do you see this developing?

Michael: As a festival we have an online community of over 3,000 members,[10] and we engage and share tips and opportunities around the smartphone filmmaking and the film industry. We also organize meetup at international festivals[11] around the world such as the Berlinale Film Festival, a gathering where smartphone filmmakers connect, share and network with each other. The community of smartphone filmmakers is growing so fast, and this is where I see smartphone film festivals are really doing remarkably well. Aside from celebrating the filmmakers, we created forums, groups, online training that help build the community. In years to come we will have most filmmakers shooting only with smartphone (Figure 3.27).

Max: Any projects or people/practitioners that inspire you?

Michael: Locally, I am inspired by the works of Kunle Afolayan, a Nigerian film director and producer; he is also a disruptor, who changed the narrative in Nigeria film industry, made a high budget film that travels to festivals all over the world and won several awards for his film *The Figurine*. Before the making of the film, Nollywood produced very low budget films made in couple of days and were distributed straight to VCD/DVD. His film reinvented Nollywood and is the starting point for the New-Nollywood. Since then the

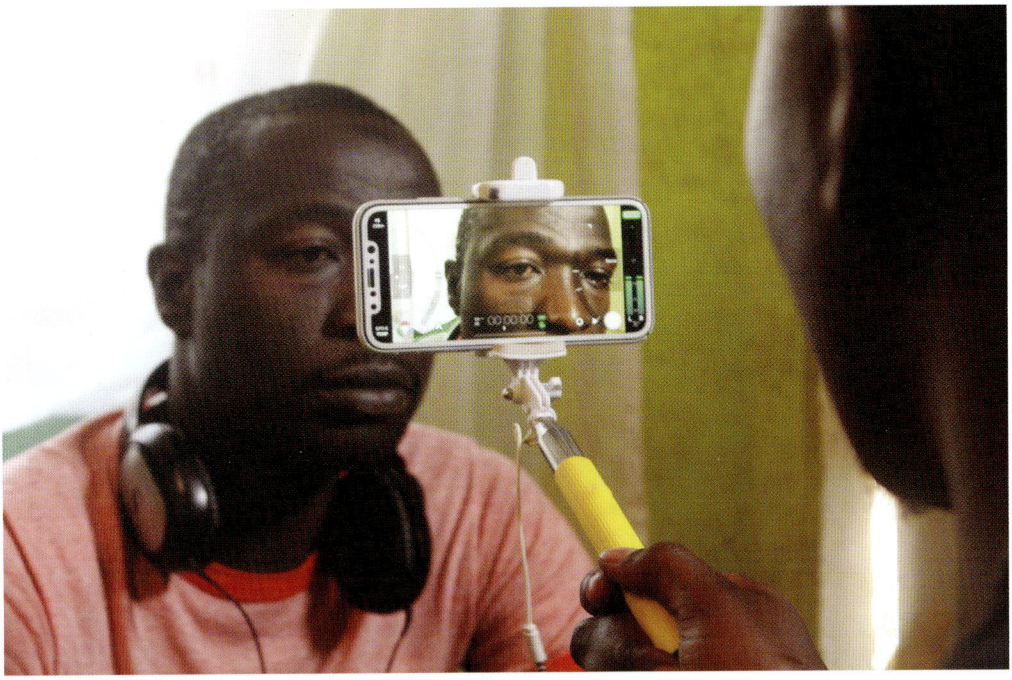

Figure 3.27 Behind-the-scenes production of *CHOCHA*, a film by Sammy Mwaura Ngetich. Won best African film at ASIFF second edition in 2018 (ASIFF 2018).

Nigeria film industry has improved in terms of budget, story, picture and sound and the theatres are now one of the major distribution outlets internationally.

Another filmmaker that inspired me locally is Niyi Akinmolayan; he is also a director from Nigeria, who has printed his name in gold in Nollywood. Niyi directed most of the blockbusters in the Nigerian cinema box office. His films *Wedding Party 2* and *Chief Daddy* were amazing, and the audience loved them. Filmmaking is also a business. Niyi is someone in Nollywood who is making producers smile to the bank.

Internationally, Tyler Perry inspired me a lot; he is one filmmaker who is true to telling original African American story, he knows his audience and he makes film for them without any compromise. He is a great filmmaker and perfect businessman.

Then, I love what Steven Soderbergh is doing with smartphone filmmaking. His transition to using iPhone to make films is amazing and has encouraged more filmmakers to follow this path.

Max: Is there anything that you would like to mention?

Michael: Yeah! Short film will be the next big thing in the nearest future, and people will get tired of watching feature-length films, and there will be huge demand for short content, which will always be available for the audience due their demand. It is what everyone is consuming on social media and is going to be on all viewing platforms including the theatres (Figure 3.28).

Smartphone filmmaking is going to take DSLR cameras out of the stores, except the companies innovate and make them into pocket devices, they might have no place in the future because smartphone filmmaking is the future (Figure 3.29).

Figure 3.28 Godwin Josiah, the Critics Company, during the production of *Z: The Beginning*, won best production design, best Nigerian film and Grand Jury Prize at ASIFF 2019 (ASIFF 2019).

Figure 3.29 Peter Oyenuga, behind-the-scenes production of *Broken*, won best Nigerian film at ASIFF first edition in 2017 (ASIFF 2017).

3.6 Karl Bardosh, NYU (New York, United States)

Max: Many thanks for being part of the book project *Smartphone Filmmaking: Theory and Practice*, and it is great to have you as part of this project because you have been one of the pioneers in that space. How did you get into organizing smartphone film festivals or back in the days mobile/cell phone film festivals?

Karl: We started very early on, actually in 2004, as one of the first ones ever. Years before the iPhone, so what happened is that I was asked to be Artistic Director for a film competition called Digi 24 in Hungary by two friends, Gabor Kindl and Laszlo Nagy. It was in Hungary but international in nature. Our filmmakers used digital cameras and made three-minute films in 24 hours and that is why it was called Digi 24. We started that format and it became very popular and later on there were 48-hour competitions or 72-hour competitions all over the world. I was thinking, well we are using digital cameras so the phone is also becoming digital so why only cameras and not the phone. As the artistic director of the competition I said let us introduce shooting with camera phones as well. In those days, iPhone wasn't even on the map. The leading phone company around the world was the Finnish phone company Nokia, so most films were shot on Nokia at that time. We launched the competition in 2004. By 2008, I made it into a global competition with contributions from the Philippines, Australia, Brazil, the United States, Africa and Europe.

In 2007, I introduced mobile filmmaking in India, and it was specifically a festival for cell phone cinema, working in collaboration with Sandeep Marwah, the president and owner of the Asian Academy of Film and Television, which is the largest and foremost media education institution in India (Figure 3.30). We started with a US/India coproduction of three narrative short films, which I wrote and directed. As the president of Marwah Studios in Film City, Noida, Sandeep provided a professional crew consisting of thirty members including lighting and dollies, but the films were shot on a Nokia phone by myself.

Figure 3.30 Professor Karl Bardosh of the NYU Film School announces Indo-American coproduction of first cell phone short films with Sandeep Marwah, Executive Director of the Asian Academy of Film and Television, at a Press Conference. Film City, Noida, 17 January 2007 (Bardosh 2007).

In 2008, Sandeep Marwah started the International Festival of Cell Phone Cinema[12] in India and now in 2019 we are at the eleventh year of that. In 2009, I got my Cell Phone Cinema course accepted at NYU, the first of its kind in the United States if not in the world and it is in its tenth anniversary now. Forward to May 2019, I took part in two competitions during the annual Cannes Film Festival – not as part of the official festival but through the Marche du Cinema's Global Short Film Awards, which have been accepting short films up to 15 minutes from all over the world. I have started the Mobile Phone Filmmaking category in this festival of short films and the best ones from different countries were awarded during the closing night of the festival in the most gorgeous Carlton International Hotel in Cannes.

Another Cell Phone Cinema completion I have started as part of the Global Peace and Tolerance Initiative run by Princess Angelique Monet for eight years within the Marche du Film of Cannes. Most exciting for me as a pioneer of Mobile Phone Cinema, that award was established in my name: it is called the Karl Bardosh Humanitarian Cellphone Cinema Award, meaning that the short films have to be on humanitarian subjects (Figure 3.31).

Max: Do you think that smartphone film festivals are somehow different from traditional industry film festivals?

Karl: The answer is yes and no. The no means there is no difference in terms of picture quality and no difference in technical quality. And many of these festivals that I described

Figure 3.31 President of the first Humanitarian Cell Phone Cinema Awards, the Hollywood director Lawrence Foldes who is in the Executive Committee of the Student Oscars, president of Jury & Artistic Director Karl Bardosh (Cell Phone Cinema Pioneer) and AFI World Peace Initiative's founder HRH Princess Angelique 'Ademiposi' Monet. This award was handled out by Princess Angelique Monet's World Peace and Tolerance Institute at the American Pavilion in Cannes as part of the Marche du Film (Bardosh 2019).

that I am organizing, the technical quality is very high. But I know other festivals in Hong Kong,[13] Toronto and San Diego among others also insist on high quality (Figure 3.32).

The more filmmakers push the envelope, the more they succeed, and we know that one of our former students from NYU, Sean Baker, two and half years ago made an outstanding film called *Tangerine*.

He made the film for one hundred thousand dollars in Hollywood using four iPhones but with a converter he was able to attach anamorphic lenses to shoot the film. Technically speaking, the phones today can record 4K and that is double the level of films used to be because the films in the theatres were mostly 2K. Smartphones can now record 4K. Sean attached an anamorphic lens and ended up with a 4K picture in Panavision format. And his film was acquired at the Sundance Film Festival and distributed not only in art house picture theatres but in commercial theatres. And the one hundred-thousand-dollar budget film became hugely successful commercially (Figure 3.33).

The largest Indian film festival in the United States is called the New York Indian Film Festival, and it has been running for some twelve or fourteen years now. We created a section in the festival that is called NYU Mobile Bollywood, but it means that I give my

Figure 3.32 TSFF – Toronto Smartphone Film Festival 2019. From left to right: Paula Ner Dormiendo and Scott Morris (both winners of TSFF 2019 Best Canadian Smartphone Film); Matt Joyce (winner of Best Performance at the 2019 48-Hour Smartphone Film Challenge); Sharonah Masson (Executive Director of Raindance Canada); Divya Chand, Cherelle Higgins and Yusuf Begg (winners of Grand Prize and Best Director at the 2019 48-Hour Smartphone Film Challenge); Mingu Kim (Festival Director of TSFF); Waleed Abuzaid and Babatunde Ajani (winners of Best Cinematography and Best Sound at the 2019 48-Hour Smartphone Film Challenge); Francisco Perez (winner of TSFF 2019 Audience Choice Award) and Jennifer Ye Won Kim (Creative Strategist and Host of TSFF) (TSFF 2019).

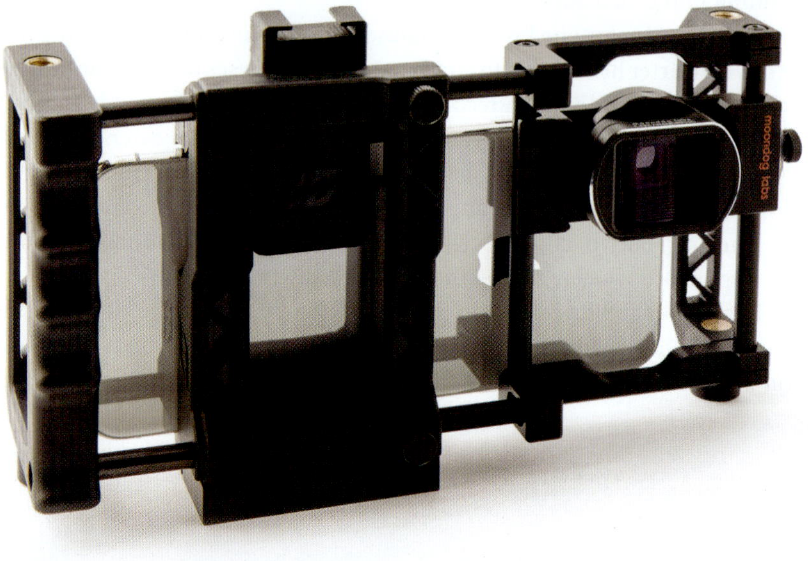

Figure 3.33 Anamorphic lens for smartphone. The film *Tangerine*, also mentioned on page p. 2, 52 and 195, as well as the smartphone feature film *Blue Moon* (see p. 10, 84, 94, 215 and 216), used the Moondog Labs set-up. The picture displays the Moondog Labs lens on a Beastgrip rig. (Moondog Labs 2019).

students in my course an assignment to shoot a one-minute music video that is cut to a Bollywood soundtrack.

Even a few years ago, when the films were screened at the festival, everyone including the Bollywood stars asked me: 'Professor are you telling us that these films that we watched on the big screens were shot by phones?' I said: 'yes.' They said: 'but we can't tell the difference' and I said, 'but that is the point'. On the technical level, that is where we made huge progression with the new smart phones.

With *Tangerine* of course nobody could tell the difference as it was shot with four iPhones with anamorphic lenses in Panavision format. Then Stephen Soderbergh, the great Hollywood director, came out two years ago with a feature film called *Unsane* and followed now with a second one and he even said that from now on he will always shoot on phones with stars using regular Hollywood productions.

So, the answer to your question: Is there a real difference between cellphone cinema and regular film? I would love to say, No. But that is not the full story, because there are certain aspects of cell phone cinema that are different from traditional cinema. One major difference is the aspect ratio because with the cell phones you can shoot in vertical format, but you cannot do that in regular cinema, it's all horizontal frame format. In fact, you can move over to a vertical, horizontal back and forth and can even start a shot in vertical and slowly turn it to horizontal in the same film. The other difference that makes mobile the most popular in the world is its short format. You can work in the long format but the most popular form in the current popular usage is the short format.

Max: With all the technical and creative developments that happened in the last five to ten years in the world of mobile, smartphone and mobile filmmaking, can you think about one thing that really excites you for the next five to ten years?

Karl: The reason why I got into cell phone cinema very early on, really among the very first pioneers, was because as a film educator, I realized that this is a device that you can use to teach and make films, edit on them, complete and even distribute them to certain media outlets through the internet; it is a full studio technically. That is why I have subtitled my course as *Hollywood in Your Pocket*. But the most important element is that I, as film educator, am interested in smartphone filmmaking because I realized that I can teach the language of film to a large number of people around the world and get a more educated audience. Obviously, if the audience feels that they can shoot films, they can edit, they can become a more sophisticated audience, they will know how to speak film language and they will learn the language of film and that is why I got interested in this, in the first place.

There is one more that goes beyond all of this, and that is that the smartphone is the cheapest computer that poor people can afford all around the world and that can connect them to the internet and also to each other. That can make them join together electronically, as Marshall McLuhan predicted in a 'Global Village' (1962). This is the number one device that turns us into living in a global village that brings us together from around the world even with poor people and offering a service that otherwise would not be available which is saving lives around the world. It is the most important tool and so again referring to McLuhan, he also said that the medium is the message. What that means is that this medium, the cell phone or smartphone, has radically changed our lives.

Max: How do you engage with the community of smartphone filmmaking and how do you see this community developing?

Karl: In my festivals like the Global Short Film Awards, the best film that won in 2019 was by Nina Zaretskaya, a professional filmmaker from Russia who made an absolutely beautiful, very touching portrait of a Holocaust survivor in a five-minute cell phone film. Really touching. She is a professional and she has her own professional company. Cell phone is a big equalizer, and anybody can join so it is not limited to amateurs or students; many professionals participate, and they are welcome.

Max: Do you feel that there is a community of smartphone filmmakers?

Karl: Yeah, probably but not really organized worldwide. There are some that congregate only at festivals but there are several festivals specifically devoted to this, and I believe that could be at least half a dozen. These are outstanding festivals some of them which I keep in touch with like the San Diego International Mobile Film Festival and obviously the Cell Phone Film Festival in Hungary and in India and China. But there are also other film festivals for specific smartphone filmmaking in Hong Kong, Toronto and Denmark such as the Dogma Festival and of course the Mobile World Conference in Spain in Barcelona. Talking about the unity, these are all filmmakers who participate and congregate around these festivals and they are quite sizeable.

Max: Have you noticed any filmmakers that are maintaining a focus on mobile, pocket and smartphone filmmaking?

Karl: Stephen Soderbergh's decision that he will shoot feature films only on cell phones from now on is not only the best example of what you are asking about but, in my opinion, it goes beyond filmmaking. The shooting and editing on the phone do not have to lead to professional filmmaking career to be successful. In fact, in my class at NYU, which is not taught in the film school but in the Open Arts School, which means that my students come from all over the university from different disciplines. It could be somebody who will become a doctor, the other actor, the third could be a biologist. Then I tell them that today there are no publications in any field of study without video. Video is becoming more and more accessible (Figure 3.34).

Max: You are recognizing the fact that smartphone filmmaking has become part of twenty-first-century creativity and with the universal language of video and online dissemination, there is a democratization of filmmaking in the process I would not see as being reversible. Are there any projects, people or practitioners that inspire you?

Karl: This question has been asked by the Insider Magazine just over a month ago that I should name my influences, my favourite films and filmmakers, and I was really hard-pressed to answer that. Scholastically, but still briefly I would have to be looking at the history of cinema and look at the different waves of filmmaking, national waves, like the mid-sixties, late sixties like the giants of cinema Fellini, Bergman and Bunuel. And then there are the French New Waves, very few people know the great Czech New Wave was that preceded the Russian invasion of the country in August 1968 and then that resulted in some Czech filmmakers coming to the United States who then became my greatest filmmakers like Milos Forman, a multiple genius Oscar winner who passed away not long ago and so currently if you want, I don't know whether to just call it a wave or a group maybe a combination of the filmmakers group because right now the Oscars are ruled by a group of Mexican filmmakers in the last couple of years. They win the Oscars one after the other and Iñárritu won it twice in a short period of time, which has never happened.

Figure 3.34 Karl Bardosh presenting at the TEDx Lecture at New York University about his Cell Phone Cinema Class, which has been running since 2009 in NYU Tisch School of the Arts (Bardosh 2009).

Max: Once there will be an Oscar-winning film produced on a smartphone. Next year will be ten years of MINA.

Karl: That is fantastic to say that you ran ten years of cell phone cinema in Australasia. It is a huge milestone and congratulations on that. We should celebrate it in Cannes.

Notes

1. Film festival submission platform FilmFreeway, https://filmfreeway.com/.
2. SPB Podcast Mobile Film, https://podcasts.apple.com/us/podcast/sbp-podcast-mobile-filmmaking/id1296673665.
3. Mobile Motion crowdfunding campaign via Kickstarter and now via Patreon, https://www.patreon.com/momofilmfest and https://www.kickstarter.com/projects/thirdcontact/help-us-keep-mobile-motion-film-festival-moving?ref=discovery&term=Mobile%20Motion. https://www.kickstarter.com/projects/14933638/help-mobile-motion-film-festival-move-further?ref=discovery&term=Mobile%20Motion.
4. Karl Bardosh Cell Phone Cinema Humanitarian Award, https://www.filmfestivals.com/blog/cannes_market_dailies/afi_world_peace_initiative_cannes_karl_bardosh_cell_phone_cinema.
5. Smartphone camera phone, https://momofilmfest.com/samsung-unveils-108-megapixel-smartphone-camera-sensor/.
6. MoMo as part of Top 100 film festivals on FilmFreeway, https://filmfreeway.com/festivals/curated/top-100-best-reviewed-festivals.
7. *Silent Eye* series on Amazon Prime, https://www.amazon.com/SilentEye/dp/B07F73QN3J.
8. ASIFF listed among the world's best smartphone festival on FilmFreeway, https://filmfreeway.com/africansmartphonefilmfestival.
9. Viral news story about emerging Nigerian filmmakers who shoot sci-fi movie with a smartphone, https://www.africanews.com/2019/08/14/nigerian-teens-make-sci-fi-films-with-smartphones and https://www.youtube.com/watch?v=D25NrAQZLng.
10. ASIFF's online community on Facebook, https://www.facebook.com/africansmartphonefilmfestival/.
11. Connecting international smartphone filmmakers with meetups such as Berlinale 2018, https://www.facebook.com/africansmartphonefilmfestival/photos/a.411120472576607/784151228606861/?type=3&eid=ARDvWmcATUCM0ZBWJwc3qFloE5AV1o7_eKt5Ht_wEThU6vPogLHFgfw9L8anGxA65BDRacbgfZo_-B6e&__tn__=EEHH-R.
12. International Festival of Cellphone Cinema in India, http://ifcpc.com/.
13. Hong Kong Mobile Film Festival, http://mobilefilm.hk/2015/en/index.php.

References

Anderson, Chris. 2004. 'The Long Tail', *Wired Magazine*. https://www.wired.com/2004/10/tail/.

Crespi-Valbona, Montserrat and Richards, Greg. 2007. 'The Meaning of Cultural Festivals: Stakeholder Perspectives in Catalunya', *International Journal of Cultural Policy*, 13 (1): 103–22.

Derrett, Ros. 2003. 'Festivals & Regional Destinations: How Festivals Demonstrate a Sense of Community & Place', *Rural Society*, 13 (1): 35–53.

Derrett, Ros. 2004. 'Festivals, Events and the Destination', in Ian Yeoman, Martin Robertson, Jane Ali-Knight, Siobhan Drummond and Una McMahon-Beattie (eds), *Festival and Events Management: An International Arts and Culture Perspective*, 32–51. Amsterdam: Elsevier Butterworth-Heinemann.

Duarte, Fernanda and de Souza e Silva, Adriana. 2014. 'Arte.mov, Mobilefest and the Emergence of a Mobile Culture in Brazil', in Gerhard Goggin and Larissa Hjorth (eds), *The Routledge Companion to Mobile Media*, 206–16. New York: Routledge.

Kim, Soyoung. 1998. '"Cine-Mania" or Cinephilia: Film Festivals and the Identity Question', *UTS Review: Cultural Studies and New Writing*, 4 (2): 174–87. Reprinted (2005) in Chi-Yun Shin, and Julian Stringer (eds), *New Korean Cinema*, 79–91. New York: New York University Press.

Labourdette, Benoît. 2010. 'Pocket Film Festival Catalogues'. *Festival Pocket Films*. https://www.benoitlabourdette.com/actions-culturelles-et-pedagogiques/conception-d-evenements-culturels/festival-pocket-films/catalogues-de-toutes-les-editions-du-festival-pocket-films?lang=en.

McLuhan, Marshall. 1962. *The Gutenberg Galaxy*. Toronto: University of Toronto Press.

Schleser, Max. 2014. 'Connecting Through Mobile Autobiographies: Self-Reflective Mobile Filmmaking, Self-Representation and Selfies', in Marsha Berry and Max Schleser (eds), *Mobile Media Making in an Age of Smartphones*, 148–58. London: Palgrave Macmillan.

Schleser, Max. 2014. 'A Decade of Mobile Moving-Image Practice', in Gerhard Goggin and Larissa Hjorth (eds), *The Routledge Companion to Mobile Media*, 157–71. Routledge: New York.

Schleser, Max, Wilson, Gavin and Keep, Dean. 2013. 'Small Screen and Big Screen: Mobile Film-Making in Australasia', *Ubiquity: The Journal of Pervasive Media*, 2 (1): 118–31. https://doi.org/10.1386/ubiq.2.1-2.118_1.

Simons, Jan. 2011. 'Between iPhone and YouTube: Movies on the Move?', in Geert Lovink and Rachel Somers Miles (eds), *Moving Images beyond YouTube*, 95–107. Amsterdam: Institute of Network Cultures.

Wilson, Gavin. 2014. 'Cell/ular Cinema: Individuated Production, Public Sharing and Mobile Phone Film Exhibition'. PhD Thesis, University of Leeds, Leeds. http://etheses.whiterose.ac.uk/8475/.

Wilson, Gavin, 2014. 'South Korean Film Festivals for Mobile Cinema: Sites for Cultural Translation or Vehicles for Segyehwa?', *JOMEC Journal* (6).

Films

Aunt Tillie's Kitchen. 2018. Dir. Mickey Harrison. United States: LuLu Films.

Blue Moon. 2018. Dir. Stefen Harris. New Zealand: Dark Horse Films.

Charlie. 2018. Dir. Miranda June Mullings. United States: independent production.

Chief Daddy. 2018. Dir. Niyi Akinmolayan. Nigeria: Ebonylife Films.

Departure. 2015. Dir. Ruben Kazantsev. United States: Belarus and France, independent production.

Dingleberry. 2017. Dir. Mickey Harrison. United States: independent production.
Express. 2017. Dir. Brian Hennings. Australia: Lot 3 Media.
Focus. 2018. Dir. Brian Hennings. Australia: Lot 3 Media.
Frankenstorm. 2012. Dir. Max Schleser. United States: independent production.
Good Day to Die. 2016. Dir. Adrian Jeffs. Australia: independent production.
High Flying Bird. 2019. Dir. Steven Soderbergh. United States: Extension 765, Harper Road Films and Netflix.
King Kong. 2005. Dir. by Peter Jackson. New Zealand: WingNut Films and Universal Pictures.
Lord of the Rings. 2001. Dir. Peter Jackson. New Zealand: New Line Cinema.
Max with a Keitai. 2006. Dir. Max Schleser. Japan: independent production.
Silent Eye. 2018, 2019, 2020. Dir. Simon Horrocks. UK: London Production Studios and Mobile Motion Pictures.
Tangerine. 2015. Dir. Sean Baker. United States: Magnolia Pictures.
The Bullying Problem. 2017. Dir. Miranda June Mullings. United States: independent production.
The Figurine. 2009. Dir. Kunle Afolayan. Nigeria: Golden Effects Studios and Jungle Film Works.
The Hobbit. 2012, 2013, 2014. Dir. Peter Jackson. New Zealand: New Line Cinema and Warner Bros. Pictures.
Third Contact. 2013. Dir. Simon Horrocks. UK: BodyDouble Productions.
Unsane. 2018. Dir. Steven Soderbergh. United States: Regency Enterprises, Extension 765 and 20th Century Fox.
Wedding Party 2. 2017. Dir. Niyi Akinmolayan. Nigeria: Ebonylife Films and FilmOne.

4

Focus on smartphone filmmaking practice

Chapter Outline

The first article that engaged with the creative potential of mobile media is the 'Aesthetics of Mobile Media Art' (Baker, Schleser and Molga 2008). Writing in collaboration with Camille Baker, who is interviewed in this chapter, this article outlined the mobile-specific creative practices. It explored three distinctive Creative Arts approaches in documentary film, mobile performance and interactive art and noted the similarities when working with mobile devices. This article defined characteristics of intimate and immediate mobile media art qualities, which are still resonating in contemporary smartphone films, as outlined in Sections 2.6 and 6.4. It expands further on studio and industry filmmaking roles and related production workflows, and explores the creative process in the realm of moving-image arts. Nowadays smartphones can produce the work in a Hollywood format and simultaneously open up new opportunities. Filmmakers and creatives that expand the boundaries do not receive the mainstream recognition their work deserves for their engagement with smartphone filmmaking. In sync with the first chapter, filmmakers and artists were chosen whose work was screened at the International Mobile Innovation Screening or presented their smartphone films and creative projects at the MINA Mobile Creativity and Mobile Innovation Symposia.

4.1 Independent smartphone filmmakers

Gerda Cammaer's experimental smartphone films screened at MINA in 2014, 2015, 2016 and 2020. She is a Belgian scholar and educator who lives and works in Toronto, Canada. She also contributed to the International Mobile Innovation Screening acting as a member of the MINA Screening Committee and co-curator for *Shifting Boundaries: Noteworthy Mobile-Mentaries (Mobile Documentaries) of the Four First International Mobile Innovation Screenings (2011-2015)*[1] at the Visible Evidence XXII conference in Toronto 2015. Her iPad films contributed to the definition of the poetic and experimental mode in Section 2.6.1. Her travel films and documentary art apply experimental film practices, which reflect mobile media's affordance of mobility.

Felipe Cardona is a filmmaker from Columbia working and living in the United States. His filmmaking style references Soviet montage aesthetics that he developed into a collaborative audiovisual format. His smartphone films can be situated at the intersection between poetic and experimental as well as participatory and engagement mode. His work also references the mobile-specific theme of 'being on the move'. Some of his smartphone films are also edited on mobile devices and thus are relevant for the smartphone native mode – such as *Bogocali*, which is a fusion of time and space as much as documentary, or rather mobile-mentary, and music video (Figure 4.1). Other mobile moving-image works are *éNfasis* or *North in South*. For the latter video-loop mobile-mentary, Schleser created a response, *South in North*, which documents Felipe's filmmaking approach.[2] More recently, Felipe experimented with online and interactive documentaries, such as *DespARcha2* and *DespARchand0* (Figure 4.2). He is a valued member of the MINA community and presented his work in New Zealand in 2010 and was featured in several screenings and contributed to MINA symposia via online presentations.

Camille Baker also participated online at the Mobile Innovation and Mobile Creativity Symposia and presented her Creative Arts research. She is a Canadian artist-performer, researcher and curator working with mobile media art and various other art forms, such as participatory performance, interactive art, tech fashion/soft circuits and DIY electronics.

Figure 4.1 Still image: *Bogocali* (Cardona 2016).

Figure 4.2 Still image: *DespARchand0* (Cardona 2018).

She works and lives in London and is a Reader at the University of the Arts. Her research into *Mobile Media Performance* (Baker 2017) contributes to the understanding of the smartphone native mode, demonstrating some future directions smartphone filmmaking could take, considering current developments towards transmedia storytelling and wearable technologies. Spanish independent filmmaker Conrad Mess was chosen to represent the 'red carpet in your pocket' mode. He demonstrates that smartphone filmmaking allows one to recreate cinematic formats and storytelling formulas. He is well known in the world of short film festivals internationally and received numerous awards for his smartphone films.

4.2 Gerda Cammaer (Canada/Belgium)

Max: Thank you very much for been part of this project – *Smartphone Filmmaking: Theory and Practice*. It is great to interview you as a filmmaker whose smartphone films screened several times at the International Mobile Innovation Screenings and as an active participant in the MINA symposia and invited keynote speaker. If it is fine with you, I will just dive in straight with the first question?

Gerda: Yes . . . and thank you for asking me to be part of this project.

Max: How did you get into mobile, smartphone or mobile filmmaking?

Gerda: That was when I got my very first iPad. It was not really the iPhone, it was the iPad that inspired me to start mobile filmmaking, sometime back in 2011. It was an iPad 2, and the image quality was not necessarily that great, but I was immediately fascinated by the fact that when you film on an iPad, you could see what is filmed while filming, like on a super big viewfinder. This kind of immediate and mediated experience was really exciting to me, and it even became a major source of inspiration for my work: I try to recreate that experience of filming and seeing what is filmed at the same time, and/or having that kind of higher connection with reality. Very much like our most recent AR project, *Viewfinders* (Figure 4.3).

Working in Augment Reality is in a way a very similar experience: one takes a mediated experience and brings it back into reality so that the two are combined in the same time frame. Before these iPad or AR experiences, a filmed image was always viewed in a different time and in a different place than where they were actually filmed or been made. You had to go to a cinema, or you had to go home and play a DVD or watch it on your TV. Nowadays, this can all happen in the same frame time-wise and place-wise. This is what fascinates me most about mobile cinema.

Another part that really got me inspired is that through my very first films with iPad I got involved with MINA, the Mobile Innovation Network Aotearoa, as it was called at that time (now it is the Mobile Innovation Network and Association). Through MINA I discovered a whole new world and community. Becoming involved in MINA as a keynote, but also as someone who every year helps with the judging of films, I am exposed to interesting works from all over the world. I truly appreciate the breadth of work that is out there or the possibilities that mobile cinema offers in terms of genre and styles and length. There is no prescribed recipe for what a mobile film is supposed to be, so anything is possible. When you look at the field of mobile cinema and the creativity that is embedded in it, there are so many possibilities and discoveries to make. Also, it is a very open and accepting community, which I find both inspiring and encouraging, and it actually helps me to keep on making work, no matter how small. Since I don't have the time anymore to make long format video projects, making these small mobile videos really helps me to keep my creativity alive.

Max: Are there some films and filmmakers that have inspired you?

Figure 4.3 *Viewfinders*. Cinematic AR app and travel film by Gerda Cammaer and Max Schleser (Cammaer and Schleser 2015–18).

Gerda: I always situated myself in between experimental and documentary film, and I think where I feel most comfortable is really on that fringe where documentary film becomes experimental film and vice versa. That is the area that I am really interested in, so the films that inspire me most always have a connection with reality, yet they can be very abstract in nature. I prefer films about something that is out there, non-fiction as it were. I don't have a strong affinity with films that have been scripted or that have actors. Within the context of MINA, I have discovered many films that strengthened my belief in mobile cinema as a perfect tool to reinvent the documentary. One of these early films was the 2013 short doc *Tofu Man*. It is a very simple short film: someone walks into a night shop where a guy is making Tofu and makes a film about it. That film taught me about the possibilities of mobile cinema in terms of intimacy and immediacy because the camera is not intimidating: people will just talk to you (the filmmaker) and as the audience it is an incredible experience to be invited into someone's life accordingly. The second thing that it taught me is to never take anything for granted. Everyone has a story, and it's probably an interesting story, in this case even an extraordinary one. You actually don't know what the twist is going to be or what the plot is based on, and I find that much more interesting than a plot that has been created by someone who wrote a script, with orchestrated dramatic turns, an emotional climax and a happy ending. *Tofu Man* touched me because of its profound humanity. There are other films that I discovered through MINA, but they are very different in genre. For example, MINA introduced me to the work of Anders Weberg who was one of the keynotes at the same MINA conference I was at in 2013. His work is very inspiring to me because of the particular aesthetic he has developed, what he is able to do with mobile cinema as an art form and his almost purist approach to mobile cinema (only mobile and nothing but mobile). Later, in 2014, there was a film by Ryan Fox called *Sketch-Three: Avant-Garde* (*R.P.M. 2*). He tied an iPhone to the wheel of his car and rode around into the night, all the lights creating an abstract painting of circular movements and colours. I find that a fascinating film because you really don't know exactly what you are looking at: he is filming reality and by the 'technique' used it becomes something completely imaginary and beautiful to look at. It's almost like an animation film and only at the end when the car stops you realize what you have been looking at, a revelation that takes you out of the magic of what you have just been watching.

Max: If you look at your own works, is there a commonality that you could see reappearing in them?

Gerda: At the core of all my work is that I look for some kind of connection with the world unfolding. Indirectly a lot of my work is inspired by some of the ideas of Gilles Deleuze, particularly his interpretation of the world as a body of infinite folds and surfaces that twist and weave through compressed time and space. The world is never standing still, but it is there, and you can go out and connect with it whenever you want and however deep you want. I do not like working with artificial light, or artificial elements, or as I said earlier, with actors and scripts, etc. I have more of a fascination with the extraordinary in the ordinary. A good example of this is my last film *Night Sail* (Figure 4.4), a video poem about sailing at night by full moon, or *Ultamarine*, two works that build on my passion for water and waterscapes and that are very much in tune with my very first film, shot on 16 mm, *Hydromancy*. But I have also been working with the very specific 'low resolution' look of mobile film that in a way we are losing now because smartphone cameras are getting better and better.

Figure 4.4 Still image: *Night Sail* (Cammaer 2019).

I have used this in particular for all the works I made inspired by travel and mobility, where each time the 'effects' of shooting while in movement with limited exposure adjustments and a fixed shutter speed, such as light flashes and motion blurs, were integrated, and sometimes even enhanced in the work, like in *Mobilarte*, *InTransit@NeedHelp?*, *Départ de Lille* and *Stoll (and Stumble) to the End of Things*, all works that were filmed on iPad. I'm also working on a series called *Kinetic Traveling Memories*, which combines various travel experiences such as riding a train in the UK with the experience of filming it on iPad or iPhone. So far, I have finished only one video of this series, but I hope to be able to edit the next two or three soon (Figure 4.5).

I would say the element that links all my work is that I explore movement in all its meanings and forms. I film not necessarily things that move but movement – which includes stillness and stagnation as well – and that probably sums up best everything that I have done so far as filmmaker, even what I did in 16 mm, long before switching to mobile cinema. By the way, a good example of an iPad video that is more about stillness and stagnation (in contrast to speed and movement) is *Velocity*.

Another core element in my work is sound design. I think my biggest achievement working with sound was for the film *Mobilearte*, which was shot on my first iPad while riding a tuk-tuk in Maputo (Mozambique) in 2012 (Figure 4.6). The sound of the video-recording was completely useless because the tuk-tuk was making a lot of noise, there was a lot of wind and the tuk-tuk's plastic curtains were flapping like mad, which all made the sound completely unusable. I wanted to create a soundtrack that was still linked to the images and that could recreate the feeling of the actual tuk-tuk ride. With the aid of sound designer Dafydd Hughes, we created a computer program to scan the images for the four national colours of Mozambique and then we turned these into four different tones that were activated when one of the colours appeared on screen. In other words, with these four notes we formed a musical soundtrack parallel to and derived from the images that I then combined with ambient sound to give the film some of its realism back.

Figure 4.5 Still image: *Kinetic Traveling Memories* (Cammaer 2016).

Figure 4.6 Still image: *Mobilearte* (Cammaer 2013 and 2014).

Some of my work has 'only' a musical track, often with music I have found for free online; other videos have specifically designed sound pieces and are made in close collaboration with a sound designer, as for example *InTransit@NeedHelp?* (also in collaboration with Dafydd Hughes) (Figure 4.7), and in the case of my 'pocket camera' film *Glissando* a music ensemble composed a piece of classical music specifically for it. But my favourite way of working is to combine ambient sound with ambient music, if I can find the right elements for it. I actually consider sound to be the most creative part of doing mobile cinema, but sound is also by far its biggest challenge.

Max: The next question relates to the limitations of the smartphone and can you think about a situation where these became an advantage?

Gerda: While these limitations are often what professional makers will qualify as bad images, or sounds, for experimental filmmakers these are often an important creative tool. This is how mobile, pocket and smartphone filmmaking becomes a gift, because if you really look at it, the way it works out or not, the way it plays out or not, that is often really, really beautiful in my mind.

Max: Do you have a preferred smartphone set-up in terms of gear and can you tell us a bit more, which are your favourite tools and what excites you about them?

Gerda: In general, and this might be different from other mobile filmmakers who are also experimental filmmakers or documentary filmmakers is that I don't use a tripod, I do not use a selfie stick, I do not use a monopod, I don't even use a frame or anything else to hold either my iPad or my iPhone. One of the things that I have found most important for myself as filmmaker is that I actually have the device directly in my hands and that I am closely connected to the tool itself that I can work with my body and use my body to stabilize the image or to make a certain movement, even if that means that sometimes the image is going to move more when the wind blows me over or when I am on sail boat such as in *Night Sail* and the boat tips to one side: obviously that is not going to be a stable image. But it is very important to me to have the camera very close to me and not on

Figure 4.7 Still image: *InTransit@NeedHelp?* (Cammaer 2016).

'remote' control by putting it on a tripod for example. For me, filming is first and foremost a bodily experience and that is even more the case for smartphone filmmaking.

Max: Do you think you will always stay with the smartphone camera or do you think you will shift to a different camera at a certain point in time?

Gerda: Obviously that will depend on what happens in terms of technological evolution. But I think what we should worry about most is how the industry is already trying to redefine smartphone filmmaking for us by putting in certain controls and filters that they think we as filmmakers want. The issue is that what they have in mind is a mainstream kind of cinema and that is not what I want. I will give you an example; in the beginning iMovie was an extremely interesting and useful editing program, but if you use it now, everything is pre-programmed into what they think people want and it is no longer a tool of use to us creative filmmakers, because it doesn't do what you want if that is not something that is mainstream. That is a worry that I have for the future, that more and more mainstream functions and controls will be built into smartphones and that all film gear will be pre-programmed in a certain fashion. Also, everything that is linked to surveillance capitalism, such as web-tracking, data mining, consumer profiling and targeted marketing, makes me wonder if and how they follow you when you are filming, or know what and where you are filming, and I think that is a very worrisome evolution. Whether I will stay with smartphone filmmaking depends on all those things because I am very concerned about my privacy, as much as I am concerned about preserving my personal creativity. But at this point in my career as a filmmaker I know for sure that it has always been small media and small film formats that have been most of interest to me. I never shot on 35-mm film, but I love 16-mm film, even developing my own film. I like the independence that I have as an artist to work with something that is small and that I don't need a big budget for, or big expensive technology, or a large team of people such as a big crew and several producers. Pocket cameras and smartphones are the tools right now that can help me to record and perceive the world in a very intimate way, which is really at the core of what I try to do and that I will continue to pursue as an artist.

Max: If you consider what happened since 2010 in the world of mobile and smartphone filmmaking, almost a decade ago, what can you imagine for the near future?

Gerda: One of the things that is going to happen immediately now is the acceptance of vertical cinema. Samsung recently launched a screen that you can use to watch images either horizontally or vertically. There has always been a lot of uncertainty in mobile cinema if you can turn your phone to film vertically (as one would to make portraits of people for example) which has always been dismissed by people from the industry as wrong and has been corrected and 'covered up' by bracketing the image on the screen. I think that is going to be over now, because this kind of reluctance to actually think of moving images in a different frame – notably a vertical frame or 'portrait mode' – is already much more mainstream. There is nothing wrong with people filming vertically, and I always found it kind of interesting that people just do it, right or wrong. But a lot of things need to happen now to actually make vertical cinema more accepted and possible (on TV or in cinemas for example). In terms of mobile viewing (watching moving images on a smartphone) we are already there.

A different answer to that question is that when I look back at the history of mobile filmmaking, one of the biggest revolutions is really the fact that so many people now have easy access to filmmaking and that new voices emerge. This has political consequences and it has artistic consequences, as mobile cinema has opened the door for people who otherwise would never have considered to become filmmakers. I think that is going to destabilize the mainstream film industry even more, because the monopoly that they had, or the kind of closed door, elitist, high-budget, high-profile position that they had for so long, is really crumbling. Basically, anyone can be a filmmaker and therefore film is a lot less elitist now. In the era of analogue films, it was still expensive to make a 16-mm film, there were certain technical skills required and you needed to have access to a lab for example, but even then, there were still lots of small independent and experimental filmmakers doing their own thing, just being brave and surviving in that mode. Now that film tools have become a lot more accessible that is only going to continue to happen at a larger scale.

On top of that, and I am not entirely clear yet for myself about if this is a good revolution or not, is that people watch images on their phones a lot more. Since this is on a very small screen, we need to start thinking about the screen in novel ways now. To watch a film on a small screen on your own is a very different experience than when you are in a theatre with an IMAX screen in front of you and with other people sitting around you. It is also a very different experience to watch a Virtual Reality work in a head-mounted display or using an Augment Reality app and combining moving images in that way with your surroundings.

Actually, if you look at the big picture and you look back at all the years that we have behind us with mobile cinema, we have to conclude that mobile film or smartphone filmmaking is really the catalyst and the point of reference for all these other technical and creative developments that now the film industry is interested in, but that they actually dismissed in the beginning.

I just would like to add that what I make or what I do as a filmmaker is not necessarily who I am as a teacher or what I bring to young people or other projects. For example, I have been working within the community like you do as well with *24 frames 24 hours*. I did a mobile film project called *Taxi Stories* which is all about filming while riding in a taxi and having a conversation with the taxi drivers, letting them tell their own stories about immigration, about changing countries and how their lives now all involve driving a taxi in a strange country (Figure 4.8).

More recently, I also did a mobile film workshop with immigrant and refugee women as part of *The Shoe Project*. *The Shoe Project* is a project that was initiated by Katherine Govier with the Bata Shoe Museum to help immigrant women telling and performing their personal experiences by telling a story about shoes. I introduced some of these women to mobile filmmaking and how to tell a story with your smartphone. The excitement and the growing confidence in these women were one of the biggest rewards and a huge inspiration to me. They were quickly discovering the kind of the artist or filmmaker they can be, what can happen only thanks to smartphone filmmaking, because they already had the tools in their pockets and a basic knowledge of how to use it. They just needed to believe that they can produce something that is worth watching and sharing with the world, just like they had done with the shoe stories.

For me the biggest step forward is that we can give people confidence that they can shoot beautiful images and make beautiful films – particularly for communities who traditionally felt insecure about making films, because they were intimidated either by the technology or by the budgets needed to make films. Now we can tell them: 'you can do this, and you have the device for it in your pocket.' That is why I find 'pocket cinema' an interesting term, because it makes mobile filmmaking even more tangible and doable in people's minds and available to everyone: it's in your pocket, at your fingertips, and there is no need to be in the industry's pocket.

Another thing that I would like to add is that I actually teach mobile filmmaking in the undergraduate film programme in my university, and I have called the course Microcinema and Mobile Filmmaking. I did this because I think that it is important to explain how mobile filmmaking now is linked to things that happened in the past. For example, if you look at the very beginning of cinema and what the Lumière Brothers were doing, that is

Figure 4.8 Still image: *Taxi Stories* (Cammaer 2017).

very similar: they were making very short films, just filming the everyday, their own family, some practical jokes or their travels. They were just (as) fascinated by the fact that they could record the world as is. Since then, through the history of cinema, there have been many other film movements that have ties with what we call smartphone filmmaking. I think it's important to let young people know about both how this is something new/ exciting and how it is connected to film history.

Max: The Lumière cameras were also projectors, some of the people that were part of the Lumière Brother's screening as audience also appeared in the film, such as *The Workers Leaving the Factory*. And the Lumières as Entrepreneurs send filmmakers around the world. Films were produced and the screenings were hosted almost in the same place, like today on social media.

Gerda: Plus, they filmed their diaries and even selfies too! O:)

Max: Thank you again for the interview.

Gerda: You're welcome. My pleasure.

4.3 Felipe Cardona (Columbia)

Max: How did you get into smartphone, mobile phone, iPhone filmmaking?

Felipe: Just by chance, in 2004 I wanted to show a mini DV short film called *Justicia Voyeur* at interfilm Berlin. Heinz Hermanns and Heide Schurmeier asked me if I was keen to make another short film with a Siemens mobile phone; I said yes and they sent me the SX1. After I sent the 90 seconds movie, they invited me to Berlin in November 2004, and I won that very first mobile film festival with the shortfilm *Checklist* (Figure 4.9).

Since then, I've been testing possibilities of mobile filmmaking with several mobile devices, Siemens, Nokia N series, and indeed making short films as well. In 2009 I bought my first iPhone and when in 2010 iPhone 4 and iMovie were launched, mobile filmmaking was finally truly an independent exercise, because the need of a full-featured desktop PC and professional software stopped being an imperative.

Max: Are there some films that have inspired you?

Felipe: Of course, in 2004 I was a big fan of Michel Gondry and Chris Cunningham's music videos. When mobiles and YouTube came to our lives nearby 2005, all changed, including smartphone filmmaking, and my favourite films were uploaded to YouTube such as Lasse Gjertsen's short *Amateur* and Tanaka Hideyuki's short film *Bond TV*. Surrealism and innovative sound treatment in films always moved me, and films like Erik Gandini's *Surplus* or Godfrey Reggio's *Koyaanisqatsi* are my favourites, exploring almost a non-narrative storytelling approach, at the moment interactive projects are growing in complexity and I am exploring the possibilities in their particular storytelling form, as a combination of halfway film, halfway internet. *Out of My Window* is a project that gives an idea of international collaboration, and *Hell's Pizza* YouTube web series and Netflix's *Bandersnatch* show how complex these projects can be and offer alternative pathways to explore a story (or stories).

Max: Of course every project is a bit different because you work with different people, you work on different locations, you work in different environments, but is there something that you can think about your production process that you can see in all of your films being realized?

Figure 4.9 Still image: *Checklist* (Cardona 2007).

Felipe: It is something related to sound, cutting, musicality, maybe movement. In fact, rhythm and/or repetitions are important source of ideas. No matter if the shooting or the final video is quiet, with or without music, sound is very important, even its absence in contrast with other sound sources. I consider that repetition is a valid illusion of sense, as valid as montage itself, the same illusion contained in poetry. Using smartphones for filmmaking gives to filmmakers a true freedom to look for a personal voice, a particular storytelling form, and opens a door for experimentation, not necessarily attached to a film influence, compromise or commitment, breaking rules or ignoring them.

Max: Do you think you can do some things with a smartphone that you might not be able to do with a different camera for instance? Have you considered the smartphone's limitation as an advantage for you?

Felipe: In 2004, limitations were a lot, and certainly all were advantages and novelties: low definition, a versatile and tiny capturing device (that fits in every pocket, corner, space) and problems on audio capturing (for instance) became an opportunity to build a sound design for a film from ground (foley included). In 2010, the chance to do all in one device was also an opportunity to explore more apps, software, to figure out how to deal with video files from one app to another, for instance. Of course, this only device let anyone to upload to YouTube a film to share it to anyone. As time has passed by, this only device is becoming into a full power one and is now the centre of a digital life ecosystem, complemented by tablets, computers, wireless networks and flat screens. Professional technology is more dedicated to a task at a time, a pro camera is not necessarily connected to internet to upload or stream video, or a video editing station is not mobile enough to use it as a video camera, a smartphone is light enough to do both functions and also powerful to capture in 4K, edit in 4K and upload in the same definition to a server and also versatile enough to chat with 'audience'.

Max: Do you have a preferred smartphone set-up in terms of gear? Can you tell us maybe about which are your favourite tools? And what excites you about this?

Felipe: There are so many gear configurations nowadays, even immersive external cameras, stabilizing complements, optics, mics, that all can be suitable for a mobile filmmaker. More is less, working with the limitations and finding alternative workflows and creative answers to technical limitations in an everyday environment.

I think extra gear that works the same as the mobile is unnecessary. I like to capture fast what I like, like any person in an everyday context, of course. If I want to capture a particular sound, I move away from other noisy sound sources. I like to capture normal video (1080 or 4K 30 fps) time lapse and also 240 fps slow-motion video. Also, I like bust mode and turn all those photos into a video with a variable framerate. And finally, screen capture is important when I want to show what is happening in the digital life, like a chat or a real-time map. In post-production I take the most of time, and take all time to edit, make music or look for it in Creative Commons repositories, look for additional sounds, make animation or video FX, as much only in the mobile as possible, using apps. I consider this one-tool-for-all practice challenging and a source of discoveries and possibilities.

Max: What software do you use for the post-production?

Felipe: I use the one that helps me to explore and lets me import from other apps more video clips. What I have seen in mobile post-production is that no matter what app you're using, video projects cannot last more than 3 minutes, because when it is longer than that, the project corrupts itself and is possible to lose the job done. That's why I make several less than 3 minutes projects if I'm making something longer than 3 minutes or just work as much in mobile as possible (in apps like iMovie, Luma Fusion, Animation Desk, Garageband, Photoshop Express, AR apps, Magisto, Animoto, etc.) and then move to a PC/Mac software like Hitfilm or Davinci Resolve. Open Shot or iMovie is also part of my mac software repertoire.

Max: Do you think you will always stay with smartphones, pocket cameras or the iPhone? Or do you think you will shift at one point to a different camera?

Felipe: What really matters of this practice is the freedom that mobiles give to experiment and look for your own voice, and more important, to let more and more people to appropriate audiovisual technology to create their own videos and films. It's a matter of social inclusion; it's not about technology.

Max: If you think back about what happened since 2010 in smartphone filmmaking, can you look into the future and can you imagine what could happen?

I think 'marriage' between film and internet is happening and interactive films are a result of this. The greatest potential in this development is to explore new opportunities for collaborative practices. Immersion or VR is a field with many filmmakers making experiments and appropriations; it is like a second chance of what Thomas Alva Edison did with his Kinetoscope and Kinetograph, a personal experience, staying away from reality for a moment, not shared, personal. Shared immersion can be experienced in online videogames, even since the times of *Second Life*. And now with the growth of AR, added images and information to a real-time streamed video, film and television are reinventing itself. The important thing of all is that people still make stories for a screen with motion pictures, editing and sound.

Max: Is there anything that you would like to mention in relation to your work or smartphone filmmaking?

Felipe: I consider that next step is interactivity, no matter if it is for a rectangular or vertical screen or for an immersive VR or AR experience, such as YouTube 360° video using a Google Cardboard and AR as in Google Maps live view or gaming with Nintendo's *Pokémon Go*, chasing monsters located in real landmarks. During so much time viewers have wanted to know more on what can be seen on a screen, and only until the final credits extra information is given in a really difficult way to read and to relate to a single moment of a movie. Since DVD technology, we can jump into chapters of a movie to 'read' on our own the story. Extra information and storytelling 'jumps' can be provided today easier than a couple a years ago, and I consider necessary to offer those two chances to new audiences, for instance a map of the place the shooting took place, social media profiles of people on the screen or the chance to participate in the creation of the proper film, video or television show – and of course, the chance to jump forward, jump back or start all over.

Max: The two smartphone filmmaking modes that you are working with is a combination of the poetic and experimental and participatory and engagement modes. Can you give some examples of your work?

Felipe: Yes. *Bogocali*: a confusing time between two cities, living in both and no one at all. A chance to make a halfway documentary/music video using smartphone apps for every task of the whole production.

DespARcha2: university students in Cali (Colombia) tell in monologues how to walk to a destination on a map, in a collaborative-mobile-interactive video (Figure 4.10).

DespARchand0: students from two universities in Cali (Colombia) show how to walk from point A to B in a map, in two very different educative institutions, in a collaborative-mobile-interactive video.

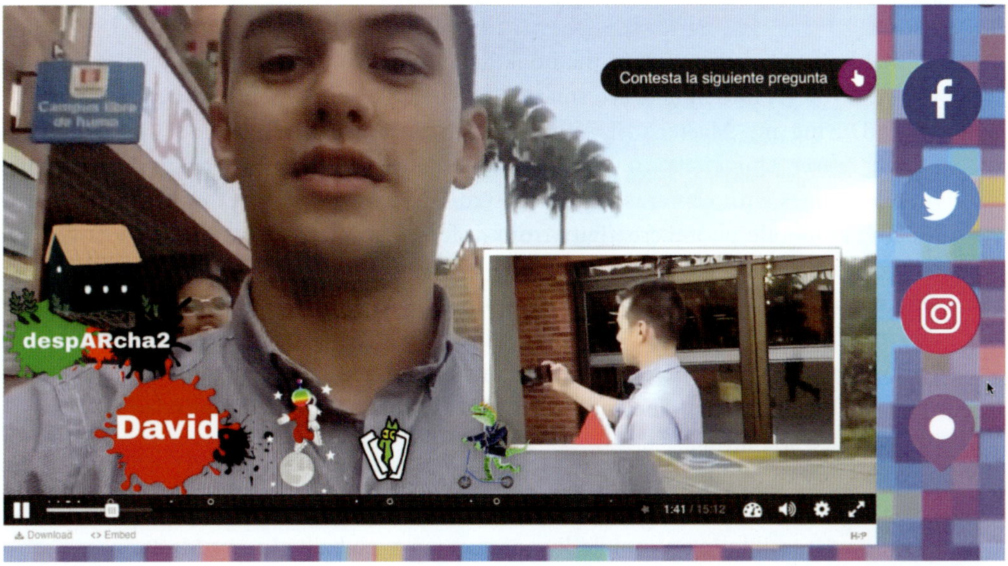

Figure 4.10 Still image: *DespARcha2* (Cardona 2019).

Figure 4.11 Still image: *North in South* (Cardona 2007).

Enfasis: a travel to Caracas (Venezuela) is also an excuse to experiment with sounds, mobile-captured video and video loops, to find the music in noise and repetition.

North in South: a journey to New Zealand full of discoveries is perfect to make video loops with video captured with a smartphone and to find out how musical could be the sounds from Wellington, Sydney, Bogota or New York (Figure 4.11).

Max: Chévere. Muchas gracias.

4.4 Camille Baker (UK)

Max: Thank you for being part of the book *Smartphone Filmmaking: Theory and Practice.* How did you get into mobile art, smartphone creativity and interactive art?

Camille: During my Master's of Applied Science in Interactive Art & Technology at Simon Fraser University in Vancouver, Canada (where I'm from), when the first mobile phones with cameras were released. There were a number of scholars already looking at mobile phone creativity from different points of view. Mainly sociological studies and some creativity, such as the networked mobile game of Ron Wakkary,[3] which look at Asian teenagers doing various things on their phone but not so much art and performance. I started thinking more about the potential of the mobile as a performance tool even before it really had video, I thought, 'Oh would it not be cool if we could have a telematic performance.' I guess you would call it now teleconferencing or live video streaming.

Max: Which is part of the *MINDtouch* project? (Figure 4.12)

Camille: The *MINDtouch* project started with the idea of connecting performers and participants, facilitating a mobile live performance globally. In my Master's research I worked with physiological sensors, bio-sensors and installation. During my Master's

Figure 4.12 Still image: *MINDtouch* (Baker 2011).

studies I was also a research assistant on Thecla Schiphorst and Susan Kozel's wearable art performance research project called *whisper[s]*. I moved to the UK, and my PhD research was conducted at SmartLab in London, which at that time was located at the University of East London. It was based on this concept of exploring mobile creativity and performance collectively live, globally, while being funded through BBC R&D to explore physiological sensing (Figure 4.13).

At that time, mobile video and streaming technology was not ready yet, the first iPhone was released in 2006, and it didn't have a camera. I used Nokia camera phones N93[4] and N95. I had to acquire a few camera phones for my mobile performance research workshops which I was conducting as most people didn't have them yet. I would ask people to explore making a video, based on their perceived physical sensations or emotions at the time – visualizing these through the video images they collected on their phones, using their available environments for visual fodder. I collected those mobile videos and uploaded them into a database and then used an early version of QuartzComposer[5] to access the videos into a live VJ a/v mix, sending it back to the internet (Figure 4.14).

The goal was to find a way to use the wearable devices on the body to collect the physiological data from each sensor and then to mix the workshop videos from the database, like a VJ. The database was categorized based on different senses (breath, muscle activity, skin condition, sweat, heartbeat). I was trying to enable people to VJ, the video

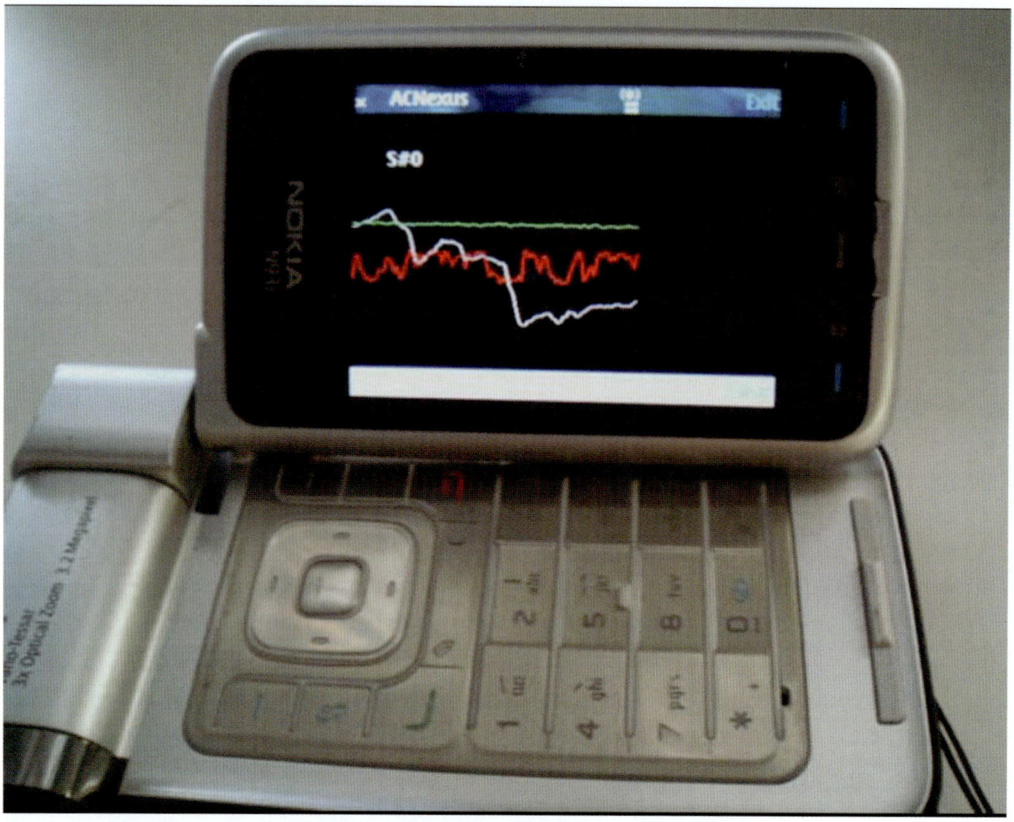

Figure 4.13 Production still: *MINDtouch* (Baker 2011).

interpretations of their body sensations, directly from their own body data, then share that performative live visual mix, globally, collaboratively with others.

I worked with a number of engineers to create workarounds to make it work, since some of the tools were not yet possible/available, at least on a localized level, so that we could perform the VJing side of it, with all of the videos in this database. Then on a global level, we could then stream the mix back out to the internet, where people could find them through their phones, which were just starting to have internet browsers installed on them and WiFi capability (though not available everywhere) (Figure 4.15).

Max: Are there any digital or interactive art projects that inspired you?

Camille: I was really inspired by the ideas from seminal new media theorist Lev Manovich[6] and Giselle Beiguelman. Manovich was exploring the idea of using the database film and algorithmic or generative 'film jockey-ing' in his project *Soft Cinema*[7] around 2003. I was also really fascinated with the concept of telepathy and how can we send images to each other mentally – not just images but feelings, sensations and thoughts. There were many artists who inspired me, like Mark America, and the whole live cinema culture of the time.

Now, I am quite inspired by some of the VR[8] performance work, and AR projects, that artists and creative technologists are now further developing. I come from a performance background, and I am really interested in how to create participatory performances using wearable and immersive technology. It is almost ten years since I completed my PhD and

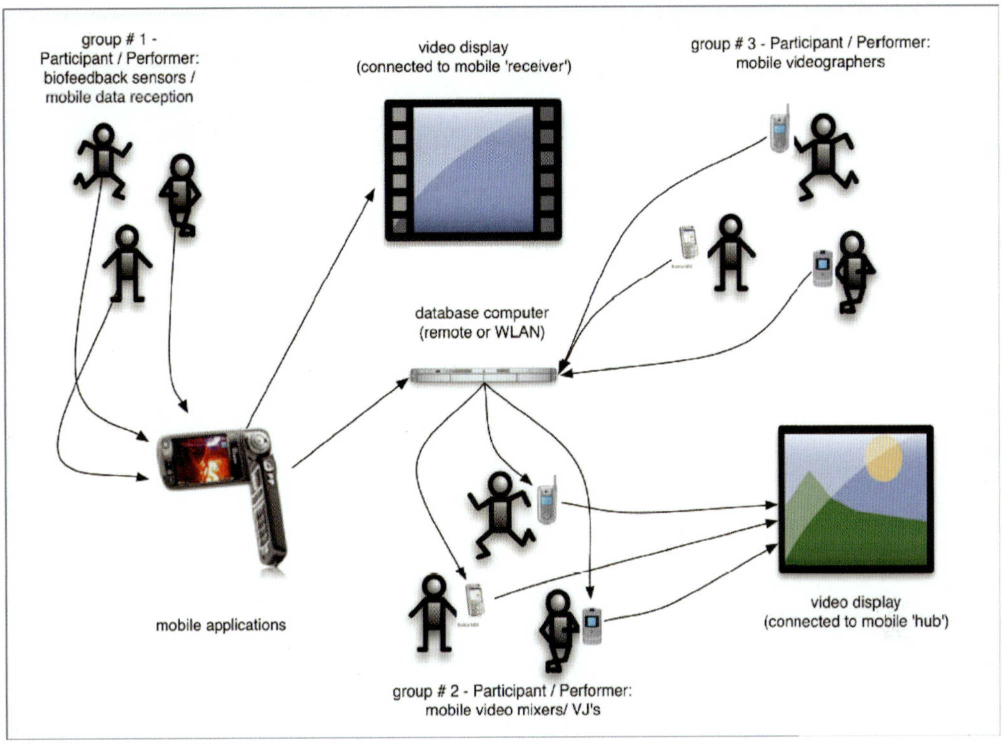

Figure 4.14 Production still: *MINDtouch* (Baker 2011).

now I am trying to find some resources and time to revisit the concepts of *MINDtouch* around live, simulated telepathic, global collaborative visual performance, using more sophisticated haptic sensing wearable technologies, enhancing a shared, immersive XR[9] performance experiences.

The experiences that have inspired me the most have been immersive theatre. When I wrote *New Directions in Mobile Media and Performance*, I researched a number of artists who in particular were trying to find ways to use a mobile phone, or sometimes a combination of wearable and mobile, with some kind of non-static screen-based, or not even screen-based, digital devices to connect with the performance. The works that I have liked so far have been mainly participatory, immersive theatre.

One inspirational project is called *Seeing I*.[10] Performance artist Mark Farid[11] creates performance work where he is wearing a VR headset for a month. Each day, he sees through the eyes of someone else, thus the name *Seeing I*. Each participant wears a GoPro head-mounted camera for a day. Farid watches the live stream through the VR headset while they were living that day of their life. I feel a connection to this work, because it is similar to my thoughts around being able to experience somebody else's thoughts or somebody else's experiences.

There is a London immersive theatre company called Dot Dot, which had been incorporating a lot of VR and immersive experiences into their theatre performances. Not always VR, sometimes it is just an immersive/planetarium-type dome environment, but they have also been incorporating headsets into their theatre pieces. One of their latest works, of the last three immersive pieces, is *Jeff Wayne's War of the Worlds* (on

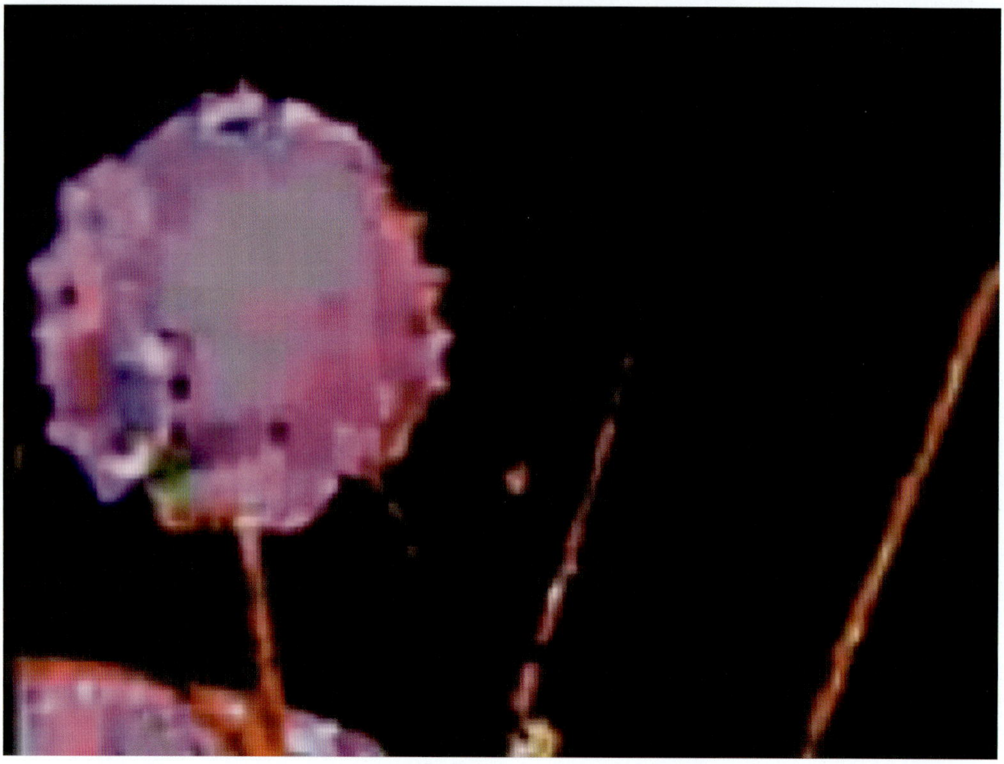

Figure 4.15 Still image: *MINDtouch* (Baker 2011).

extended run to Christmas since April 2019 due to popularity). It is a very commercial, high production-value and volume audience mixed reality experience that features many theatrical and immersive spaces, including around three that involve wearing a VR headset, to tell the popular tale from a first-person perspective.

Max: Is there an element or characteristic that you see as quite unique or specific to smartphones and technology in your creative process?

Camille: I am interested in the possibilities of the haptics and wearable/smart textiles and how to push the boundaries of the technology. I think about what is the mobile tool good for, but also how can it be emotional, to touch somebody on a deeper, even visceral level?

At the moment, I am moving slightly away from wearables and getting more into immersive experiences. I have just secured a little funding from my university for a new project that is quite personal. It is connected to my own health issue that happened a few years ago. But it is focused on creating an immersive environment that engages participants to share stories of their own health experience. *INTER/her* is focused on women who are going through post-reproductive health issues, because there are a lot of physical pain issues for women who have gone past childbearing years experiencing a range of gynaecological health concerns, and this will hopefully shed light on them and open the conversation up more publicly – problems like endometriosis, fibroids, cysts, different kinds of gynaecological cancers and menopause, problems that just do not get a lot of attention or discussed publicly, due to social shame, stigma or misunderstanding and ageism about the bodies of women over forty.

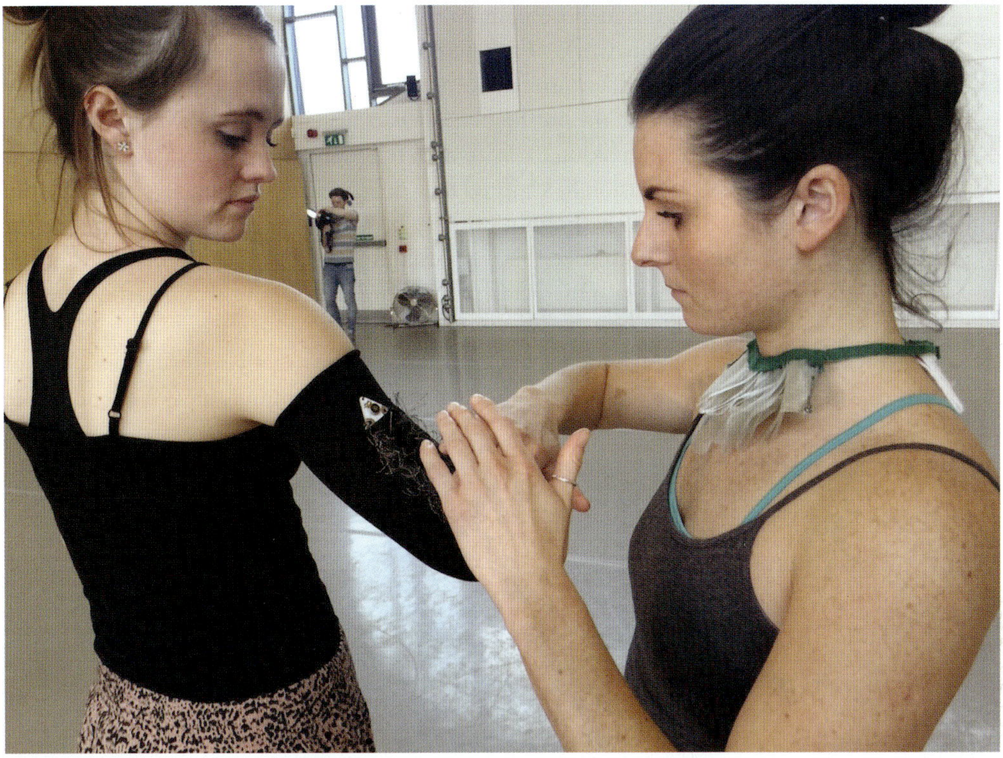

Figure 4.16 Production still: *Hacking the Body* (Baker and Sicchio 2011).

The experience will be edu-art, but also personal and empathetic. There will be wearable corset that participants will put on before entering the immersive space that will vibrate on part of the lower abdomen whenever there are stories by women discussing the pain and experience of that part of the reproductive system. It is partly really personal, and partly about helping women share experiences and learn more about their bodies and the experiences shouldn't be suffered in silence. Men can also try the experience, in order to understand their partner's and women's health more directly.

I will be making an effort to find the balance between educational and artistic. Currently, I am trying to find the right environment to show it to make it a shared experience. The mobile component is focused on the sharing of their stories. I am just working out some of the details and specifics; the project will be physicalized, felt in the body and on the body, as a shared experience. Two things that are really important to me are the shared or common and emotional elements: the visceral feeling and physical sensation.

I have been drawn to the mobile since the early 2000s, because it is so intimate and versatile, and continues to be so.

Max: Smartphone filmmaking has demonstrated that its limitations can be recognized as an advantage over standard broadcasting equipment for personal storytelling. In the context of interactive and digital art, which affordances of the smartphone are most interesting for your work?

Camille: Definitely mobility. I see the camera and the audio components as the critical pieces, besides the communication elements. I have always been interested in how we can turn smartphones into multipurpose computing tools. The intimacy and that sense of connection are also key in my work – the sense of connection in a really immediate way. Connection, intimacy, immediacy, mobility.

There is something about the intimacy that can also make it very serendipitous, which has also ended up making it invasive. The current era of increased online surveillance and privacy invasion has ignited new practices and approaches, that initially were not on my mind. I naively thought initially only of all of the positive aspects, but now I have been working more with these issues and exploring them in my *Hacking the Body* collaborative projects with media artist/choreographer Kate Sicchio from 2011 to 2016 (Figure 4.16 and Figure 4.17).

Now we are aware of a lot of the negative elements or the negative affordances. I do not like that the big corporate tech companies who make the phones have embedded the

Figure 4.17 Production still: *Hacking the Body* (Baker and Sicchio 2011).

phones with trackers – even if you do not turn these on, they are in there, and you generally have to work hard to turn them off. The internal accelerometers, and gyroscopes, alongside the various step counter apps, and so on . . . critiquing these is now a big part of the work that I have been doing of late: looking at wearables and other kinds of biofeedback sensors and how they surveil us, from a critical perspective, which was a big part of a now finished EU project I led, called WEAR Sustain, on ethical and sustainable e-textiles and wearable technologies (www.wearsustain.eu)

Max: You said you are now working more with immersive media. Do you have a preferred production workflow?

Camille: I would say no, because technology evolves so quickly, and it depends on what I am making. What I am interested in, strangely enough, is going back to my PhD work *MINDtouch* as mentioned, getting back to the concept of live streaming, but using live socially generated VR (Virtual Reality), alongside or working with haptics, to deliver the most real and immediate experience.

I use a Google Cardboard as well as the mobile HoloKit,[12] a cardboard version of HoloLens and now Oculus. I like to prototype on the low-end side of tech, see how it works and then try to implement the outcome using the high-end technology like the Microsoft HoloLens.[13] I think that's kind of how I work, trying to piece things together from a cheap side, partly because I often don't have the funding to do it high end, which is a pretty standard way of working. Then use my prototype to work towards the next level. Then that is when I usually bring in other engineers, or designers, or coders into the mix. The tech landscape is shifting so much in this area and it is hard to even keep up. It is easier to use the cheap tools to get a sense of what you want to do and then six months down the road, something else has come out.

In my view, there is a disconnect between what the big companies like Oculus, Microsoft, Google and Nintendo are making, both in terms of games, experiences and kit , and in terms of what artists are doing and exploring. And now Google is pulling out of the AR and VR headset space[14] and that is not a good sign. But then, of course, Google had no success with Google Glass.

In the project I'm trying to get funded and started, *ACTIVATE*, I am developing and exploring a first-person experience of being caught in extreme weather, without participants actually having to be in that extreme weather, to motivate people to act on climate change and support resilience for it. It is mixing the haptic with the theatrical, with the 'real' world. There is a bit of a mix of filmmaking and animation. It seems that the Unity game-engine is the software in particular that brings these all together. It is not really the same as editing or post-production. It is definitely more in the domain of an interactive experience design, almost a game-making approach, as opposed to a filmmaking approach. We are bringing assets into an interactive environment as opposed to editing, in a time-based, linear film experience, much more spatial.

Max: If you think about what happened since 2010, the last nine years, then if you just look into the next three or five years, what is the thing that excites you most about this?

Camille: Making interactive work that is responsive and generative, using smartphones or a combination of mobile, VR and wearable technology. A lot of wearable tools and technology is controlled through the mobile even Oculous Quest. Kate Sicchio and I, with engineer Becky Stewart and designer Tara Baoth Mooney, made a couple wearable tech/e-textile performance projects as part of the *Hacking the Body 2.0* collaboration,

Figure 4.18 Production still: *Hacking the Body* (Baker and Sicchio 2011).

called *Flutter/Stutter* and *Feel Me* in 2015–16 (Figure 4.18). We had an iPad controller app designed for the project, but so much is happening with machine learning and AI now, things have advanced massively in the last four to five years. Similar to the games environment, there is a lot of development by games and VR design companies into AI to interact more deeply with games[15] and make more human-like, non-playing character to interact with.

On a more general level, I see works that are really abstract, theatrical or game-focused in the coming years, more and more creatives combining non-linearity and discovery-based, spatial narratives. I think theatre companies are the best at this, because they have been excelling at audience engagement, and especially in immersive theatre[16] for years now, and they know how to tell a story from a completely non-linear discovery level and have been doing so much longer than tech or even game companies.

Immersive theatre companies like PunchDrunk,[17] and others have created spaces for people to explore and create their own narrative through a combination of theatre and narrative engagement techniques. I mean, for example, Dot Dot's *Jeff Wayne's War of the Worlds*[18] is engaging because they took a well-known story and put you inside that world and had you discover it from your own perspective, in a very personalized way (six to eight people in a group at a time yet thousands have gone to it), and they call it Layered Reality™.[19] A lot of these 'Layer Reality' type projects by Dot Dot, PunchDrunk, Shunt, Secret Cinema and others do the same thing: they put you inside of a well-known story, but now with VR and AR as well, you start by knowing a bit about the story already, then you are able to discover and reconstruct it in a way that it is still consistent with the original story.

Yet I think there is a lot of work to be done on what constitutes a good engaging narrative – how is VR different from other forms of interactive and digital kinds of narrative? I do think that it is not going to be the big tech and game companies who will come up with the most interesting work. From the works that I have seen, like I said, the most engaging are either based on a well-known story or really abstract or surreal: there is a return to surrealist age like there was for film, and now with VR and AR, and immersive experiences, because surrealism is so much easier to experiment with than narrative when you are trying out new technologies, such as at the We Are Now Festival in 2016[20] or the recent production by Marshmallow Laser Feast[21] at the Saatchi Gallery in London in spring of 2019, called *We Live in an Ocean of Air*,[22] which was a bit of an homage to Char Davies[23] Ephemere and Osmose in my view.

Max: That is a great closing point. Is there anything that you wanted to talk about, anything that I have forgotten to ask?

Camille: The only thought I would have is, at least in my work, and where I see things going is adding the haptic/feeling/sensing/dimension. Elon Musk released his full-body Tesla gaming suit[24] in 2019, so it is clear that haptics are becoming more and more part of the immersive experience, whether it is a gimmick, I do not know. I am not sure of the future of mobile media, really. Mobile is going to mean something different, I think, in ten years. It will be the merging of wearable and mobile embedded into or affixed to the body.

Then by adding the haptic dimension to enable people to feeling something, we move towards the Star Trek Holodeck.[25] It is not just because Janet Murray came up with this idea over twenty years ago in 1997. I think it is really that we want to have an alternate but realistic experience to our lives. We want to make it more and more realistic and possibly even more extreme yet safe, which is what my *ACTIVATE* project is exploring. I just see that this is where things are going. I am not sure of the future of the mobile phone or smartphone, per se, or what it will look like, but I think it is going to transform radically.

Max: Devices are getting smaller, and smaller, and smaller. Even with the iPhone, you have got the Apple Watch as a wearable component.

Camille: Exactly. Also, controllable through Siri and other kinds of digital assistants. It could be that you end up filming something through your GoPro, and it is being controlled through your digital assistant. These are the kinds of things that I think are probably going to merge.

Max: And this could include drones and robots, as a further development of Gimbals. The future is now. And of course many thanks for being a part of *Smartphone Filmmaking: Theory and Practice*.

Camille's post-interview postscript: since the Covid-19 lockdown it seems more people, artists and others will be looking for new ways to connect in real time with each other virtually, and I've discovered using the mobile-based Oculus Quest the crazy world on VRchat and AltSpaceVR and all the VR world design options that open up this conversation for new directions in my work, watch this space . . .

4.5 Conrad Mess (Spain)

Conrad: My name is Conrad Mess. I'm what is called an iPhone filmmaker. Actually, I shouldn't say this, but I'm the world's most awarded iPhone filmmaker. That is an honour for me.

Max: That is one of the reasons I was really keen to talk to you. At the moment thirty-five film festival awards? Thirty awards?

Conrad: Almost fifty.

Max: I stopped counting some time ago. Fifty, amazing.

Conrad: Fifty awards and I stopped counting at 120 nominations.

Max: Wonderful.

Conrad: Yes, it is an honour and a pleasure.

Max: How did you get into smartphone, mobile phone, iPhone filmmaking?

Conrad: It's something that I always wanted to be, I wanted to be a filmmaker since I had memories, but things of life you never take the step and make a movie. I think it was 2010 when I decided to get a camera, the holiday camera, the handy camera and get some friends and make a short film with no budget, no gears, no nothing. I guess you know I never went to film school, so I did this, it came from inside of me without a script, without a storyboard, without nothing. Just me and my camera and a few friends.

It looked like the people liked it so I wanted to make another movie but I realized that if you don't have resources, if you don't have a budget it's very hard to compete with people that have all of this. In that moment, it was 2011, and smartphone film festivals appeared in the United States. I decided to make a movie with an iPhone, with my wife's iPhone. Because I felt that it would be easier to compete with people with the same resources and budget than me.

The same thing, I shot it without a script, without a budget, just a cellphone and a few friends again, the same friends and I shot the picture. It is the most awarded iPhone short film now and so far (Figure 4.19). Sorry, I have to say that most of the nominations and part of the awards have been won comparing against real camera short films. I mean there are not 120 smartphone film festivals. Most of my nominations have been won in regular film festivals.

Max: Awesome. Are there some films that have inspired you?

Conrad: One thing is that what I want to do and the other thing is what I can do, okay? In terms of ideas or narrative I like a lot Tarantino and I think that you can see that in my first movies.

Max: Yes.

Conrad: And Robert Rodriguez. Then I also love movies from Michael Bay, I know that people hate Michael Bay but I like the style – not the narrative or how he creates the characters but the visual style.

Max: Of course every project is a bit different because you work with different people, you work on different locations, you work in different environments but is there something that you can think about your production process, in terms of when you get together with a crew, when you go on set, is there something that you have a feeling that you can see in all of your films being realized?

Figure 4.19 Still image: *The Fixer* (Mess 2012).

Conrad: The feeling common in every one of my shootings is that we do not have a budget. There is one thing that you have to understand when making a movie with an iPhone. Take your time in preparation because everything is going to go bad, everything will be wrong in some moment and you have to have everything under control and know how to react to the things that happen. You can, I think but I don't recommend it, go to the city and see what happens.

I've been learning all these years, and I realized that you have to make a great pre-production and brief your crew, three or four people is the most, my biggest crew was five people in *Time to Pay Off Debts* but everyone has to know everything about what is going to happen in the shooting (Figure 4.20). It will make you save time because a smartphone is not a camera; it is a cellphone with a lens. You have to know all the limitations that this device has.

Max: That is a great point. You talk about the limitations; do you think you can do some things with a smartphone that you might not be able to do with a different camera for instance? Have you come across some things where actually the limitation was an advantage for you?

Conrad: Yeah. We have two sides, okay? One side, if you want to go cheaper you have to go with the smartphone because it's cheap, it's easy, you don't have to know anything about cameras. That's my point, now I know a bit more about cameras, something like shutter or ISO.

I didn't know anything, exposure, I didn't know nothing about it. If you don't know anything about this, you have to get your smartphone and shoot it. Cheaper, easy, you just have to push the red button and start shooting but on the other side you can do the same things you can do with a camera because with all the gears you have in the market today you can get your cell phone and put it on a dolly or a tripod or you can put a lens on it or you can link it to an external monitor for preview (Figure 4.21). You have both sides, the easy and the professional one.

Max: Talking about smartphone filmmaking gear, do you have a preference for filmmaking?

Figure 4.20 Production still: *Time to Pay Off Debts* (Mess 2014).

Figure 4.21 Production still: *Time to Pay Off Debts* (Mess 2016).

Figure 4.22 Still image: *Time to Pay Off Debts* (Mess 2016).

Conrad: Yes, I always use 'smart phocus', it's the evolution of the iPhone.

You can attach lenses; you can attach even the Nikon or Canon ones. The thing is that they discontinued the production, so it's not available anymore. In my last movie *Time to Pay Off Debts* I used iPro lens and I liked it (Figure 4.22).

It is the only thing that I use for shooting, just a case and the lens. This is for sure, but what else? I like to use everything that I can get. I used a Steadicam for one shot in the room or a dolly. Everything that I can get access to.

Max: Let us talk also a bit about post-production, because we have talked a bit about pre-production, production, being on the set, working with friends, working with time planning, using different gear on production. But of course, if one looks at your latest films we can see that you have also worked with quite a bit in post-production. It seems you used some green screen compositing with the story. Can you tell us maybe about which are your favourite tools and techniques? And what excites you about this?

By the way I think it is quite amazing to see that one can take smartphone footage into a post-production workflow.

Conrad: First of all, I have to say that obviously from my first movie *The Fixer* to my last movie *Time to Pay Off Debts*, I have been meeting people. In that last movie I had actors, a small group, five people, they are friends, but they are also industry professionals. One thing is my first movie and the other thing is my last movie. I have been improving my work and improving my crew. I always work with my friends too, I mean I have friends from the beginning, from my first movie that I had that are now helping me in the last movie (Figure 4.23).

Having said this I used a green screen in *The Fixer* and I used green screen in *Time to Pay Off Debts*. You can use green screen even if you don't have a big crew or big resources, you just need a green blanket, this is what we used. We just had to think about how to use these things and how to improve the movie. Because in *The Fixer* we did not have lights, on set we just have a reflector and the green blanket. There is one shot of the bad guys of the picture walking towards the camera and I shot this in a gym on a treadmill because I wanted the movement very smooth.

With the green blanket behind the actor on the treadmill and the iPhone on a tripod, we could use a green screen in a low-budget movie. Then we have *The Other Side*, my movie in black and white, Victorian vampire story, where everything, I mean everything was shot in a green screen (Figure 4.24). We went to an old factory, an abandoned factory, and we put a big green screen and we made everything there.

Figure 4.23 Production still: *The Fixer* (Mess 2012).

Figure 4.24 Still image: *The Other Side* (Mess 2014).

Figure 4.25 Still image: *The Other Side* (Mess 2014).

Conrad: And the *Time to Pay Off Debts*, that same thing, the window that one can see behind the girl in composed. It is a fake window, it is a green screen (Figures 4.25 and 4.26).

Max: What software do you use for the post-production? Is it Adobe Premiere Pro or Adobe After Effects?

Conrad: To make your movie look better you have to invest in great post-production work. I edit my movies with Adobe Premiere Pro but then I make the colour correction and all the visual effects in After Effects. Because After Effects allows you to make colour grading that makes the movie look like a real movie and you can fix a lot of issues such as jelly effects.

It is the best software and Adobe does not pay me anything for saying this. You can achieve to make an iPhone movie or a smartphone movie look like a blockbuster.

Max: Awesome. Any trends in smartphone filming that you find quite interesting? You might have because if you go to lots of festivals to see other work as well, have you seen anything in smartphone filmmaking that excites you or you think, 'Oh that could be great?' With the launch of every iPhone the image quality is improving. There are lots of things happening in the space.

Conrad: There is one friend of mine, Alfonso García, that makes movies that I would like to participate because they are cool movies and I would say the movies even some of them are better than mine. Actually, all the movies that I see in iPhone look better than mine for me.

The iPhone 7 has a dual camera, I think that is a very good idea because each time you want to make a close-up you have to get the cellphone and put the camera closer to your actor. In this case, it is a 2X zoom so you can be closer to your actor without moving everything. Because that is the problem, when you want to shoot a medium shot and then you want to shoot close-up, you have to move every single light, everything and that is a pain.

Max: Do you think you will always stay with smartphones, pocket cameras or iPhones? Or do you think you will shift at one point to a different camera?

Conrad: I always say iPhone because it is what I am used to but I have a project that I wanted to shoot, it was very expensive, actually we were talking with Samsung to invest in this project. Most of them were very interested in the project but at the end the person that had to say 'Give them their money' said 'No'.

I can shoot with an iPhone, with a Samsung, shoot with camera, I do not care. Any camera is fine if the budget is right.

Max: Yeah, that is fair enough, that is what I say too. Last question. If you think back about what happened since 2010 in the smartphone filmmaking, can you look into the future and can you imagine what could happen?

Conrad: Yes, I have two versions of this story.

Figure 4.26 Production still: *The Other Side* (Mess 2014).

In one of the stories, in one of the futures the smartphone filmmaking will be what I think it should be, filmmaking for people without budget and a way in which the industry can scout people who have no resources, but they have talent. I think that should be the vision for smartphone filmmaking. Like in football there are people from big clubs that are searching in Africa or Asia for people, the same thing could happen to smartphone filmmaking.

The other future that I'm scared of, because it is already happening is that people with a lot of resources, a long career and big success are using the iPhone or the smartphone to make movies to get more response and attraction.

There are two futures that I see in this industry and one of them I like a lot, it is for people that want to show what they have inside them.

Max: Muchas gracias, awesome.

Conrad: It has been a pleasure to talk to you.

Notes

1. *Shifting Boundaries*: *Noteworthy Mobile-Mentaries (Mobile Documentaries) of the Four First International Mobile Innovation Screenings (2011-2015)* at the Visible Evidence XXII conference in Toronto, https://www.visibleevidence.org/wp-content/uploads/2017/10/VEXXII.pdf.
2. *South in North* https://vimeo.com/91507686.
3. The reGossip project co-led by Wakkary was a network game about gossip and social status *c.* 2001. The multimodal network included mobile phones, personal digital assistants and desktop computing systems. See https://www.academia.edu/11527310/regossip_gossip_and_multi-user_narratives?source=swp_share. See the journal article on it by Andersen, Holopainen, and Wakkary 2001.
4. Nokia N93 phone, https://en.wikipedia.org/wiki/Nokia_N93 and https://www.gsmarena.com/nokia_n93-review-107p2.php.
5. Early visual programming environment XCode Tools software of Apple, *c.* 2008–10, https://developer.apple.com/library/archive/documentation/GraphicsImaging/Conceptual/QuartzComposerUserGuide/qc_intro/qc_intro.html.
6. Lev Manovich, https://mitpress.mit.edu/contributors/lev-thackara and https://www.egs.edu/faculty/lev-manovich.
7. *Soft Cinema* overview of the project and subsequent book documentation here, http://manovich.net/index.php/projects/soft-cinema.
8. Let's not forget this is the second wave of VR development, which had its first wave mid-1990s with the pre-eminent artwork of Char Davies, who also started the company *Softimage* in Montreal that went onto sell its name and assets to big software company Autodesk that makes 3D design software tools *Maya* and *3Ds Max*.
9. Extended reality including AR/VR headset technologies such as Hololens.
10. *Seeing I* project info here, http://www.seeing-i.co.uk/.
11. More about the work of Mark Farid, see https://www.markfarid.com/.
12. Mobile HoloKit, https://holokit.io/.
13. Microsoft HoloLens, https://www.microsoft.com/en-us/hololens.

14. Google drops Daydream, https://uk.pcmag.com/google-daydream-view/123077/google-discontinues-daydream-vr-headset-drops-support-in-pixel-4.
15. See more on AI, VR and games here, http://sitn.hms.harvard.edu/flash/2017/ai-video-games-toward-intelligent-game/, and https://www.youtube.com/channel/UCov_51F0betb6hJ6Gumxg3Q as well as https://youtu.be/7os10TB_bcY and https://www.youtube.com/watch?v=E9OyTr8rFOI.
16. Immersive Theater definition here by Stephen M. Eckert, https://contemporaryperformance.com/2017/12/09/immersive-theater/.
17. PunchDrunk is one of the most successful and possibly the first immersive theatre company, which doesn't use technology in their productions but use fully participatory and 'choose-your-own-adventure' style theatre, https://www.punchdrunk.org.uk.
18. *Jeff Wayne's War of the Worlds*, https://www.dotdot.london/the-experience/.
19. Layered Reality discussed in the *Huffington Post* by Max Thurlow, https://tinyurl.com/ss3kgqt, and the *War of the Worlds* website, https://www.dotdot.london/layered-reality-tm/ and blog https://www.dotdot.london/everything-you-need-to-know-about-layered-reality/.
20. We Are Now Festival, https://www.roundhouse.org.uk/whats-on/2016/we-are-now/.
21. Marshmallow Laser Feast, https://www.marshmallowlaserfeast.com/.
22. *We Live in an Ocean of Air*, https://www.marshmallowlaserfeast.com/experiences/ocean-of-air/ and https://arts.vive.com/uk/articles/projects/art-photography/we-live-in-an-ocean-of-air/.
23. Char Davies who can be considered the first VR artist, who founded *Softimage*, mentioned earlier, http://www.immersence.com/osmose/.
24. For full-body haptics Tesla gaming suit – to feel the gun shots, see https://teslasuit.io/.
25. Janet Murray's prediction of where cyberspace was going, https://www.simonandschuster.com/books/Hamlet-on-the-Holodeck/Janet-H-Murray/9781439136133.

References

Andersen, Kristina, Holopainen, Jussi and Wakkary, Ron. 2001. 'reGossip: Gossip and Multi-user Narratives', in Monika Fleischmann and Wolfgang Strauss (eds), *Proceedings of CAST01/Living in Mixed Realities*, 251–4. Sankt Augustin: Conference on Artistic, Cultural and Scientific Aspects of Experimental Media Spaces.

Baker, Camille. 2011. 'MINDtouch' PhD SMARTlab, University of East London.

Baker, Camille. 2018. *New Directions in Mobile Media and Performance*. New York: Routledge.

Baker, Camille, Schleser, Max, Molga, Kasia. 2009. 'Aesthetics of Mobile Media Art', *Journal of Media Practice*, 10 (2&3): 101–22. https://doi.org/10.1386/jmpr.10.2-3.101_1.

Manovich, Lev. 2005. *Soft Cinema: Navigating the Database*. Cambridge, MA: MIT Press.

Films

Amateur. 2006. Dir. Lasse Gjertsen. Norway and Italy: independent production.

Bandersnatch – A Black Mirror Event. 2018. Dir. David Slade. UK: House of Tomorrow and Netflix.

Bond TV プリンストンガ. 2007. Dir. Tanaka Hideyuki. Japan: Frame Graphics.
Checklist. 2007. Dir. Felipe Cardona. Columbia: independent production.
Bogocali. 2016. Dir. Felipe Cardona. Columbia: independent production.
Départ de Lille. 2016. Dir. Gerda Cammaer. France and Canada: independent production.
DespARcha2. 2019. Dir. Felipe Cardona. Columbia: independent production.
DespARchand0. 2018. Dir. Felipe Cardona. Columbia: independent production.
éNfasis. 2007. Dir. Felipe Cardona. Columbia: independent production.
Glissando. 2017. Dir. Gerda Cammaer. New Zealand and Canada: independent production.
Hell's Pizza's *Interactive Zombie Movie Adventure – DELIVER ME TO HELL – REAL ZOMBIES ATTACK*. 2010. New Zealand: LITTLESISTERFILMS.
Hydromancy. 1998. Dir. Gerda Cammaer. United States and Canada: independent production.
InTransit@NeedHelp? 2016. Dir. Gerda Cammaer. UK and Canada: independent production.
Justicia Voyeur. 2009. Dir. Felipe Cardona. Columbia: independent production.
Kinetic Traveling Memories. 2016. Dir. Gerda Cammaer. UK: independent production.
Koyaanisqatsi. 1982. Dir. Godfrey Reggio. United States: Institute for Reginal Education, American Zoetrope.
Mobilearte. 2013 and 2014. Dir. Gerda Cammaer. Mozambique and Canada: independent production.
Night Sail. 2019. Dir. Gerda Cammaer. Canada: independent production.
North in South ver 1 no comments 04102013. 2007. Dir. Felipe Cardona. Columbia, United States and New Zealand: independent production.
Out of My Window. 2010. Dir. Katerina Cizek. Canada: Highrise.
Sketch-Three: Avant-Garde (R.P.M. 2). 2014. Dir. Ryan Fox. United States: independent production.
South in North. 2014. Dir. Max Schleser. Columbia: independent production.
Stoll (and Stumble) to the End of Things. 2018. Dir. Gerda Cammaer. New Zealand and Canada: independent production.
Surplus: Terrorized into Being Consumers. 2003. Dir. Erik Gandini. Sweden: ATOM.
Taxi Stories. 2017. Dir. Gerda Cammaer. Canada, Beck Taxi: Ryerson University, Beck Taxi and Harbourfront Art Centre.
The Fixer. 2012. Dir. Conrad Mess. Spain: MeSSFilmMakers.
The Other Side. 2014. Dir. Conrad Mess. Spain: MeSSFilmMakers.
The Shoe Project. 2011–20. Dir. Katherine Govier. Canada.
The Workers Leaving the Factory. 1895. Dir. Louis Lumière. France: Lumière Brothers.
Time to Pay Off Debts. 2016. Dir. Conrad Mess. Spain: MeSSFilmMakers.
Tofu Man. 2014. Dir. Andrew Robb. Australia independent production.
Ultamarine. 2016. Dir. Gerda Cammaer. New Zealand: independent production.
Velocity. 2017. Dir. Gerda Cammaer. Australia and Canada: independent production.
Viewfinders. 2015–2018. Dir. Gerda Cammaer and Max Schleser. international: independent production, http://www.viewfinders.gallery/.

Projects

ACTIVATE. in production. Dir. Camille Baker. UK: UCA.
Flutter/Stutter and Feel Me. 2011–16. Dir. Camille Baker and Kate Sicchio. UK: UCA and Arts Council England.

Hacking the Body. 2011–16. Dir. Camille Baker and Kate Sicchio. UK: UCA and Arts Council England.
INTER/her. in production. Dir. Camille Baker. UK: UCA.
Jeff Wayne's War of the Worlds. 2020. Created/Dir. dotdotdot UK: dotdotdot.
Seeing I. 2020. Performed/Dir. Mark Farid. UK: arebyte Gallery and the European Media Art Platform.
We Live in an Ocean of Air. 2020. Dir. Marshallow Laser Feast. UK: Marshallow Laser Feast.

5

New horizons Creative Industries

Chapter Outline

The omnipresence of smartphones, in combination with an increasing number of the apps and a widespread recognition of transmedia storytelling as well as cross-media production, has led to disruption within the Creative Industries sector in the last decade. Since the launch of the iPhone in 2007 and the App Store in 2008, mobile marketing and digital advertising as much as mobile journalism (MOJO) are embracing new opportunities. One can also notice these developments in the film industry. As this book reveals, short and feature films are filmed on smartphones and mobile devices, and an increasing number of apps are available for each part of the production process. Smartphone filmmaking can be described as a sandbox model and engages with viewers and makers, as well as all hybrid definitions that emerged in the last years: pro-sumer (professional consumer); pro-d-user (producer and user) and generally what was once termed 'audiences'. When studying film production, one typically considers the distinctive phases of pre, production and post-production. Several apps support each phase, and social media can be used from scouting talent to film dissemination and film marketing. The linearity in production, distribution and exhibition is no longer a Fordist model but embraces network practices. The film industry has, for a long time, focused on innovation on a technical level, as revenue streams were generated in clear-cut pipelines. The internet, broadband and for this book especially relevant 5G appearing on the horizon have afforded new approaches from crowdfunding, video and streaming platforms and media-services providing new spaces next to the studio and Hollywood film industry models. Through the online distribution, niche markets and social media, smartphone filmmaking spans from the independent sector to the studio system.

5.1 Twenty-first-century creativity and storytelling

As one example of this development of twenty-first-century creativity and digital literacy, one could point at Adobe mobile-first and app strategy. The development of Adobe Spark and the launch of Adobe's software as native apps such as Photoshop and Illustrator reveal that mobile devices are the third screen not only for consumption of media content but also production. No surprise that Adobe launched Adobe Rush. As a little sister/brother of Premiere Pro this app/software is developed for vbloggers, YouTubers and smartphone filmmakers. New niches for content creation and storyteller are interacting with the increased application of video online. Treehouse is one of the first entire mobile production houses, producing any smartphone video formats from commercials to feature films (Figure 5.1).

Within the industry context smartphone filmmakers work with a number of accessories and tools such as external microphones, light and lenses to enhance the smartphone's capacity to broadcasting standards. With the development of smartphone-specific gear, such as audio equipment, camera ricks, gimbals and tripods, lenses, a new R&D area provided

Figure 5.1 Treehouse crew: Megan, Jason, Ginny, Niall, Olivia (Treehouse 2019).

Figure 5.2 ShoulderPod set-up for attaching mics and lighting – some additional requirements for broadcasting standards (ShoulderPod 2019).

new opportunities for existing companies – for example gave rise to, Rode is producing a number of dedicated microphones with smartphone connectors or new companies, such as Struman Optics or Luma Touch. The Australian company Struman Optics produces lenses and the YouShooter kit for storytellers, filmmakers and anyone with a smartphone that can include video into their day-to-day business. The opportunities range from tradeswomen and men to real estate agents to teachers. The US-based company Luma Touch develops Luma Fusion, which is a mobile editing application for iOS and Android that provides all features one would recognize in non-linear desktop editing applications. Smartphone filmmaking thus provides opportunities for entrepreneurs and creatives alike and has an impact on the economy beyond the film industry. Other companies that have made significant contribution to the world of smartphone filmmaking and are used on a day-by-day basis are FiLMiC Pro and hardware providers like ShoulderPod (Figure 5.2).

French filmmaker, mobile phone film festival organizer and educator Benoît Labourdette, as one of the key figures and pioneers in the world of mobile and smartphone filmmaking, demonstrates how smartphone filmmaking can also make a contribution to society and culture.

5.2 Luma Touch (United States)

Max: Thank you very much for being part of this book *Smartphone Filmmaking: Theory and Practice*. How did you get into the world of smartphone filmmaking and the smartphone film industry?

Terri: I started video editing back in 1988 and have been either video editing or working with development teams on video editing systems for my entire career. As I learned more about video editing, I ended up working for companies who were making video editing systems like Lightworks, Tektronix, Fast Multimedia, Pinnacle and AVID with a primary purpose of bringing my own video editing expertise to the development of the systems they were building. One of my roles working at these companies was to be a customer advocate, describing how we, as editors, want to use the software (Figure 5.3).

At AVID, Chris Demiris and I were asked to head up a team of engineers to make an iPad editor – Chris as the lead developer and myself as the lead designer and product manager. We were given free reign to do what we wanted. And with a team of six, we set off to develop the features for a touch-based editor. You rarely get that kind of freedom in a large company, so I have to give AVID credit for encouraging it at that time. What we built became the AVID Studio for iPad App, which then eventually became Pinnacle Studio for iOS. The app had a lot of great features and excellent usability, but what we could do in 2011 was limited compared to what we can do today (Figure 5.4).

It was when AVID sold the consumer team to Corel that Chris and I decided to start our own company: Luma Touch. Our first priority for Luma Touch was to expand the editing features and capabilities of an iPad or an iPhone. We created something we called the Spry Engine, which is the compositing engine that runs the LumaFusion app, even to this day. Everything for LumaFusion was built on all new code and the latest resources.

My whole professional life has been about editing, so I have seen – or been involved with – the process of conceiving and implementing new editing features. For example, if we were talking about trimming, I would probably have five or six different ways in my mind you could implement effective trimming from a user's point of view. Likewise,

Figure 5.3 Terri Morgan at Fast Multimedia in 2001 (Morgan and LumaTouch 2019).

Figure 5.4 Pinnacle Studio for iOS, April 2012 (Morgan 2012).

Chris would have five or six ways of developing it. And we've brought our best ideas and experience to developing LumaFusion. When we set out to create a feature, we don't rush through it. We ask ourselves, 'What is the most powerful, most fun, most understandable way to create this feature?' Sometimes we come up with something that is completely new and sometimes we rely on our experience from past projects to guide us (Figure 5.5).

Our apps were adopted early on by the MOJO (mobile journalist) community. Journalists like Philip Bromwell from RTÉ News, Michael Scheyer from Regio TV of Bodensee and others consistently gave us insights about what they wanted for their specific news workflows, like the ability to do audio channel mapping so they could enable independent audio levels for each channel of a stereo clip. This was important because they use two mics for recording, one for the interviewee and one for the ambient sound. It was nice to have their feedback right up front and very early on. It was real-world user feedback! (Figure 5.6)

No matter what kind of editing you are doing – news, corporate, drama or documentary – it's all about telling a story. We don't try to force people into a certain way of editing because, from our point of view, every story is unique, and every storyteller will tell a story in a different way. We try to make features that allow people to make their own decisions so that videos edited with LumaFusion are not identifiable as having come from our app. We want to enable people to edit a good story and be able to do that in whatever way they want to do it (Figure 5.7).

Max: If you reflect on what happened in the last five to ten years in the world of smartphone filmmaking, what developments are you most excited about when looking ahead into the near future?

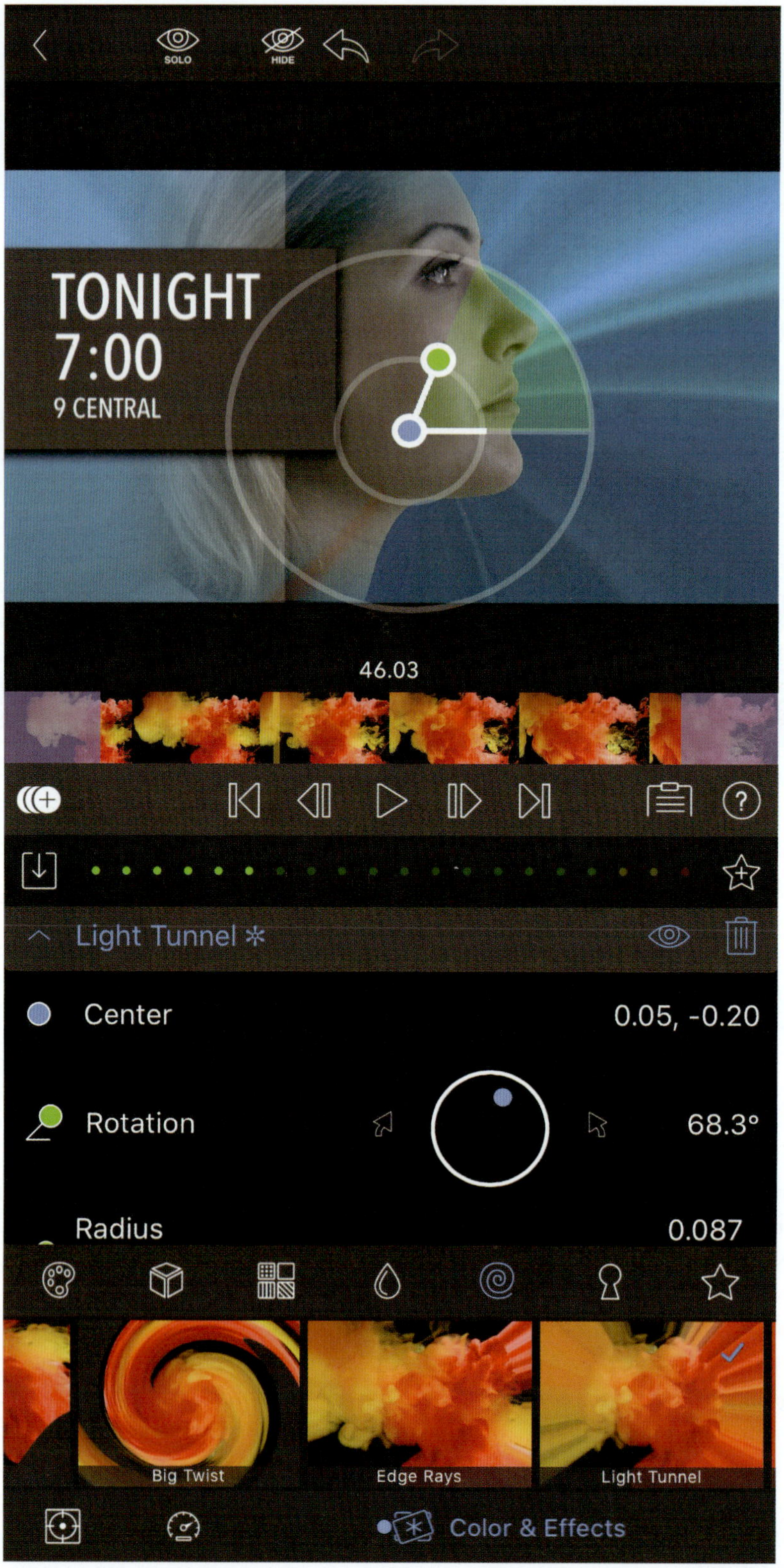

Figure 5.5 LumaFusion 2.0 Effects Editor (LumaTouch 2019).

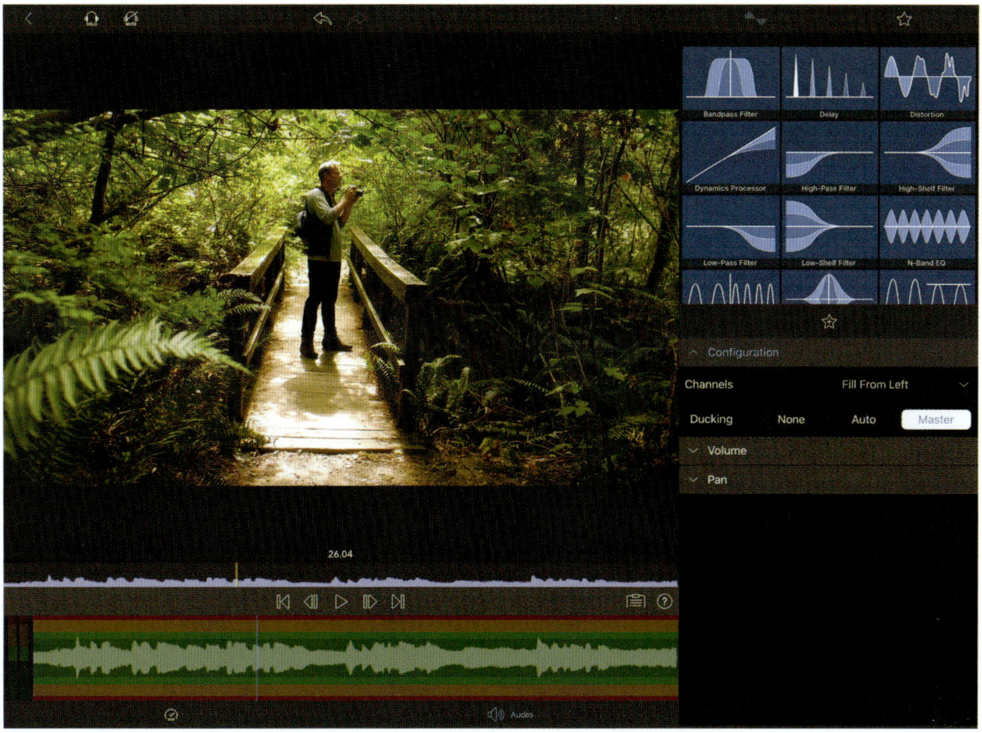

Figure 5.6 Audio channel mapping in LumaFusion (LumaTouch 2019).

Figure 5.7 LumaFusion 2.0 timeline on an iPad Pro 12.9 (LumaTouch 2019).

Terri: The past few years have proven that with smaller and smaller devices, we are able to do serious work. But I am really excited about where we are with mobile editing because during the process of developing for mobile, we've found there are some other unexpected benefits beyond the obvious mobility of the hardware (Figure 5.8).

Mobile editing changes the way you interact with your media in a number of ways. First, mobile editing liberates you from your editing studio and from sitting at a desk. You can edit when and where you are most motivated to tell your story. If you're at a café, on an airplane, in a car or lying in bed, you can pick up your device and start editing. The device turns on instantly, there's no boot-up time and no finding drives or looking for a plug-in, so it's easy to do a bit of editing and get your ideas on the timeline no matter where you are (Figure 5.9).

Editing early is another change that mobile brings. I spent years in the past sitting in an edit suite waiting while the production crew would go out, bring the media back to me, drop it on my desk with a script and say, 'Hey, put this together.' At that moment the story would be completely new to me. I would have no idea what the feeling was at the shoot, what shots they got or what they were trying to get. But with a mobile editor, I can go out on the shoot with my iPad, experience the story, put together a rough cut on-location and even give input as to what shots would make the editing better. When you edit early, you find problems early, you are more motivated and involved in the story and you are more likely to hit deadlines (Figure 5.10).

Editing is a skill that takes practice to develop. And because access to mobile editing is so easy, mobile editors edit more often. They become better at their craft for having spent more time doing it. There are times in my career where I have edited daily and the more I edit, the better I get at it. It becomes second nature to find the right place to cut. And there have been times in my career where I have edited less often because I was doing other parts of development, and when I come back to editing, I find it takes me a while to really be editing at my peak.

In the future, I think we will see more and more professionals utilizing mobile editing in their workflows and enjoying these benefits.

Max: One of the things that got me really excited when I started to work with native smartphone films; I could not only capture the moment, which resonates with ephemeral qualities, but I could also edit while I was on location. Almost like an artist that is painting with an easel on location rather than going to the studio. As a consequence, one is immersed by the environment.

Terri: I agree! It feels more joyful to edit in the moment, when you're most excited about the story. You can feel the energy of the place you are shooting at and right there you can put your story together. It is just a very fun and joyful thing to do versus coming back days later and having to put stuff together that you've kind of forgotten about. That delay before editing just kind of zaps the joy out of it for me.

Max: Do you see some potential in this area of collaborative video editing?

Terri: One of the things that we are doing right now is integrating Frame.io right into the LumaFusion library. Frame.io is an online collaborative platform for your media. It allows you to put markers and comments on your media, draw on the video and talk to other people on your team about specific frames of the video. In LumaFusion you can actually see your Frame.io media right in the library, play your media and add it to the timeline,

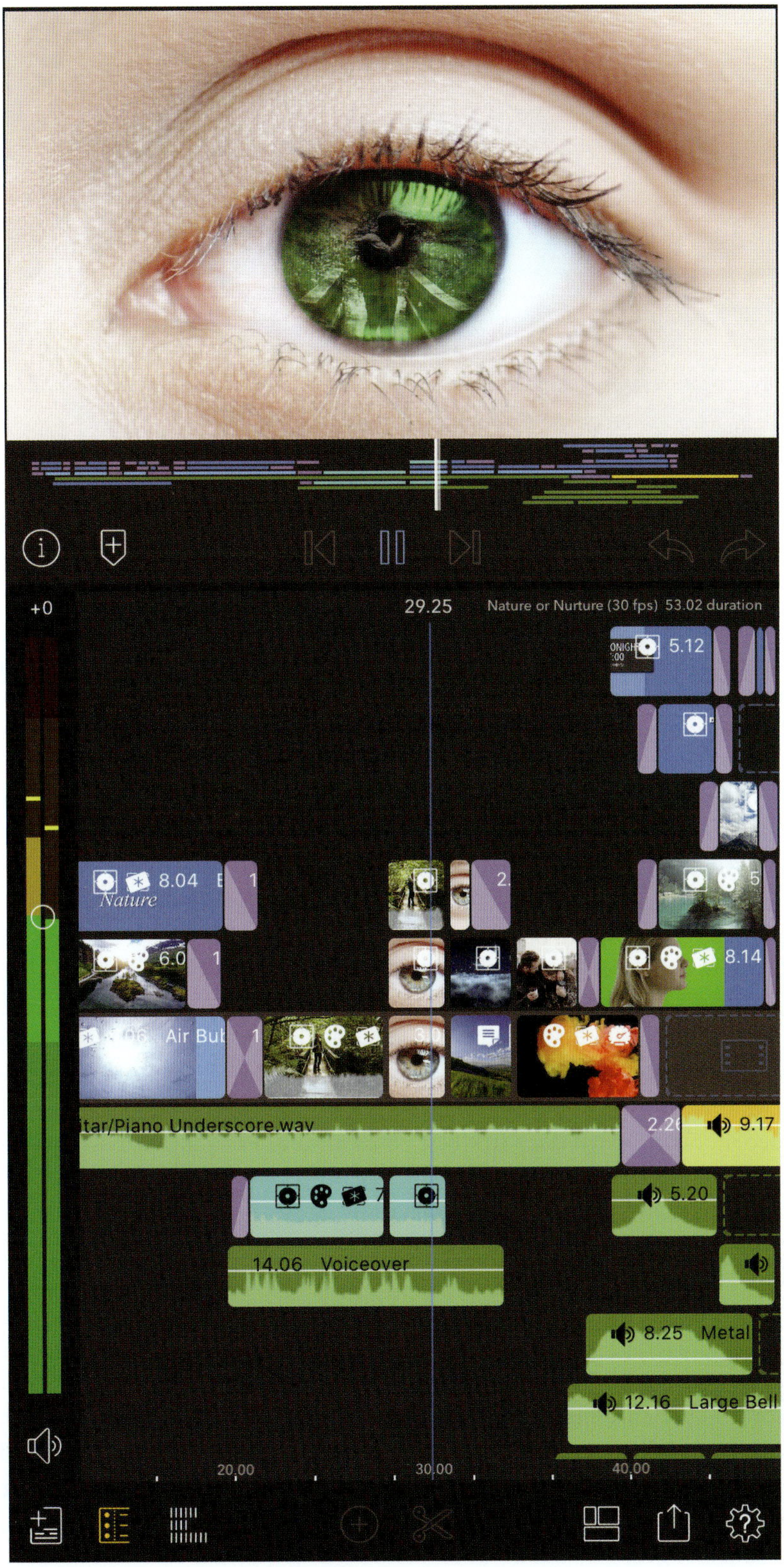

Figure 5.8 LumaFusion 2.0 timeline on iPhone XR (LumaTouch 2019).

Figure 5.9 Mobile editing anywhere (LumaTouch 2019).

and see and respond to the comments. It's a powerful workflow that we even use within our own team at Luma Touch (Figure 5.11).

Max: How do you engage with the community of smartphone filmmaking and how do you see this community developing?

Terri: We try to engage in a lot of different ways. One of the things that we have started recently is a bi-weekly web cast. It is not a high-end production; we turn on the camera to teach something about the app or something about the world of video. It is live and our whole team gets involved with the webcast. That has been really fun because people are enjoying that they can be part of the process of what we create, they can be informed about what we are working on and they can give us feedback and send ideas about things they want. It is about connecting with our customers and connecting with people who are in the industry. We also have a booth at trade shows for professional videographers, such as IBC and NAB, but there are also all sorts of mobile film festivals and conferences going on in the community like MojoFest, Mojo Italia, Rencontres de la vidéo mobile,[1] SmartFone Flick Fest,[2] SmartPhilm Fest[3] and MINA[4] that we attend or support.

The mobile filmmaking community is growing and is especially good for students who now find it affordable to create films.

Max: What are you most excited about regarding mobile editing?

When I started in the video industry, consumers were separated from professionals because the average person could not afford to shoot the same quality media that professionals could shoot. Not only was it expensive to get the equipment, it was also hard to get the experience needed to operate the equipment. Today, anyone can shoot high-quality video with just a smartphone and then continue on to do professional editing with just a smartphone or tablet. Mobile shooting and editing levels the playing field and brings us all back to the most important thing, the story. It simply comes down to who is making the best story that touches us or makes us happy, interested or motivated. That is what I am most excited about, that we can finally start measuring a successful video by

Figure 5.10 Checking we have the right shots by a bit of quick editing on the shoot (LumaTouch 2019).

the quality of the story instead of the quality of the video. With a minimum budget and a device that most people already have in their pocket they can tell a complete and compelling story.

I do have some favourite filmmakers and journalists who are doing all their editing from their phone or iPad. One journalist is Philip Bromwell from RTÉ News, who is

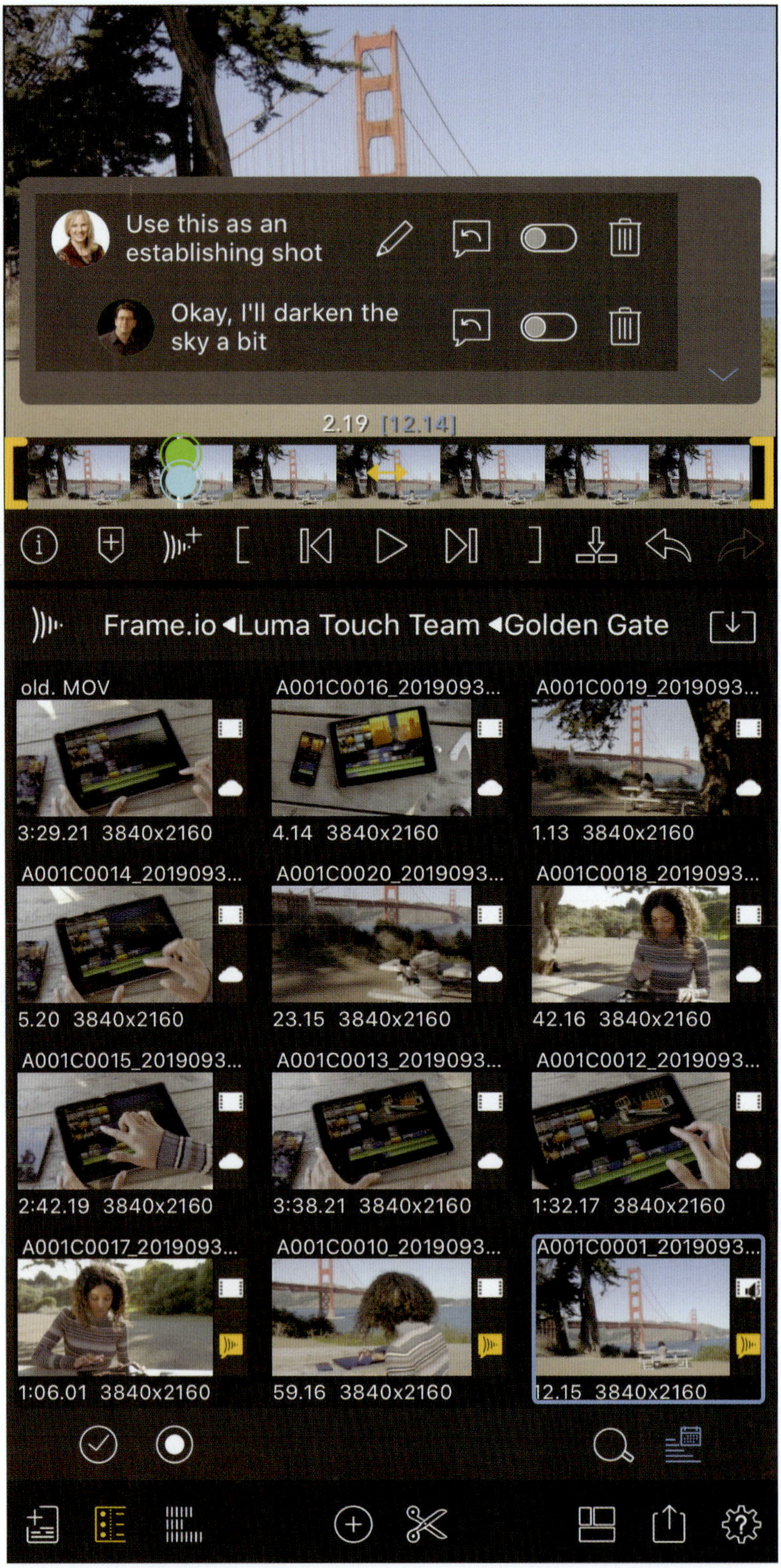

Figure 5.11 Frame.io integration with LumaFusion 2.0 (LumaTouch 2019).

creating absolutely amazing public interest stories using Filmic Pro and LumaFusion. You would never be able to tell that these were shot and edited on a smartphone. He has recreated the official RTÉ News graphics right in the LumaFusion app as well so he can upload a completed piece to the station.

There are also people like Rob Layton who teaches mobile journalism and smartphone photography, and he is an awesome underwater iPhone photographer and filmmaker (Figure 5.12). He has taken me places through his videos that absolutely I am not going to go (that is under water because I do not really like being under water). But watching his videos gives me this complete sense of submersion and again he is doing this all with his iPhone for shooting and editing (Figure 5.13).

There are endless examples of people who are taking me places I never would have gone to on my own. Angelo Chiacchio edited the *Ephemera* documentary series on iPad

Figure 5.12 Ocean iPhoneographer and mobile journalism educator Rob Layton conducts an underwater iPhone photography workshop in the pool at Bond University. Shot on iPhone 8Plus using the native camera app in AxisGo housing with flatport. Edited in Lightroom CC on iPad Pro. 1/48000sec f1.8 iso25 (Layton 2019).

Figure 5.13 Bubbles float to the surface of the ocean. Shot on iPhone XS using native camera app in AxisGo housing with dome port. Edited in Lightroom CC on iPad Pro. 1/2900sec f1.8 iso25 (Layton 2019).

with LumaFusion. Angelo spent 300 days exploring and documenting thirty of the rarest and most fragile places on Earth before they disappear. He took only what he could carry on his back. He shot with both his iPhone and a DSLR camera.

Once he was done shooting, he used LumaFusion on an iPad Pro to edit each sequence before he would move on (Figure 5.14). It was important for him to see what he was able

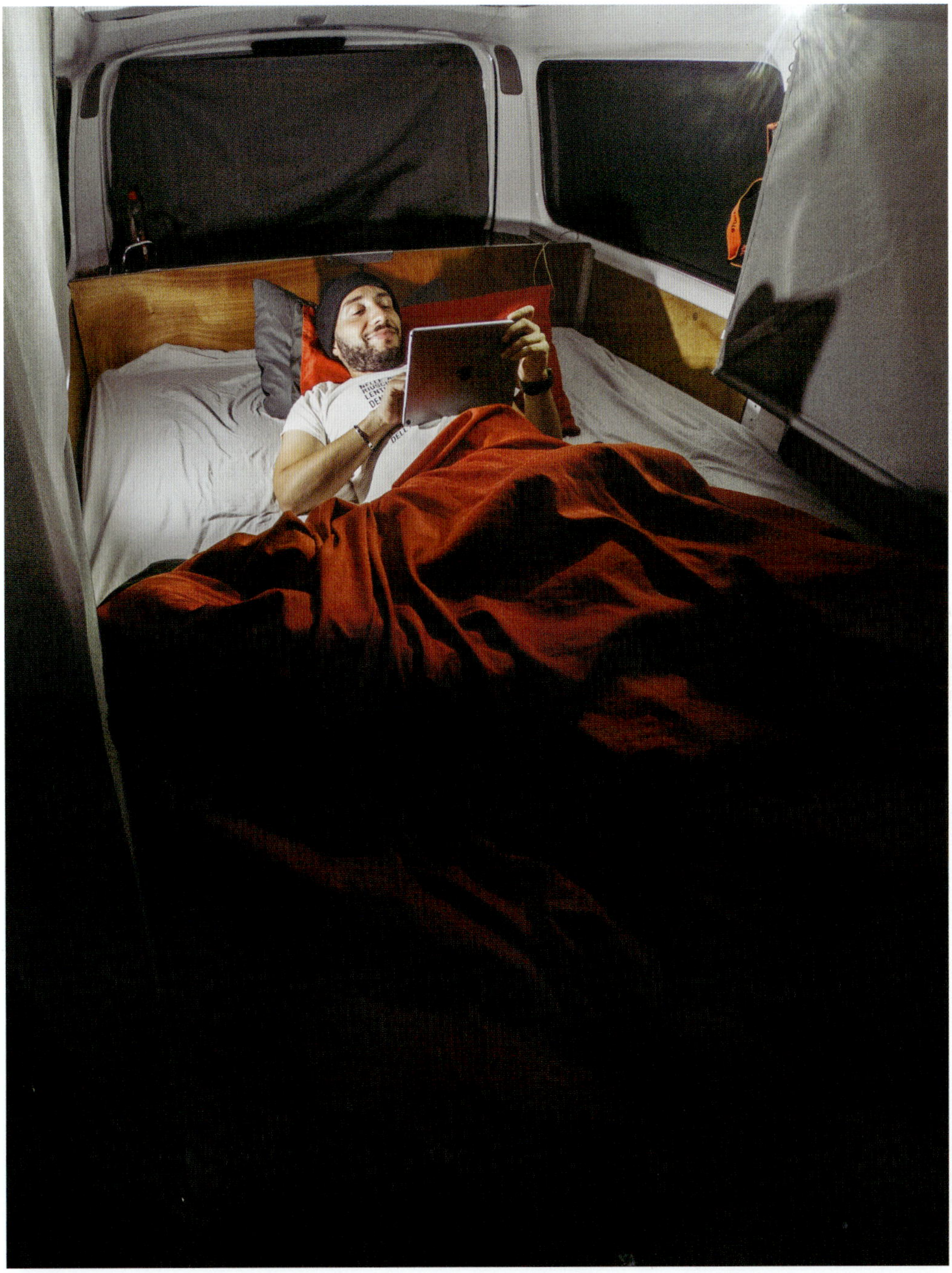

Figure 5.14 Production still: *Ephemera* (Chiacchio 2019).

to create before leaving since going back to get another shot was not a viable option. The next time he got to a location with wireless he quickly uploaded the finished movie. His documentary features some of the most beautiful and captivating stories about different cultures that I have seen[5] (Figure 5.15).

Max: Can you share some best practice examples and approaches to smartphone editing?

Terri: I have a few tips. The first one is regarding your equipment and how to manage your media. Traditionally the computer we used for editing and the computer we used for other things like mail and games were kept separate whenever possible. The reason for that was to avoid conflicts of software and avoid accidentally deleting files or losing your work. But now many people are shooting on smartphones where they also shoot all of their personal videos, answer emails, talk on the phone, text, watch videos and play games. We often put the device in our pocket or purse without protecting the lens, and we often manage our storage without much thought. It becomes difficult to keep your lens clean and protected, and even harder to keep enough free space on your device to edit a film and still have enough room on the device to export the film. Even if you are exporting to a different location like Dropbox or YouTube, the device needs to have enough room for LumaFusion to create a temporary file for upload. So, my first piece of advice is to take care of your device, not only the outside of it but also really think about your storage and where all your files live and how much storage you have. Make sure you have enough to finish your movie and export it. Consider using external drives with USB-C connections if you are using an iPad Pro. You can connect LumaFusion to the drive so you see the media right inside the LumaFusion library and LumaFusion will only transfer the clips you add to the timeline, saving you room on your device.

When you begin to edit, it can be a bit overwhelming to decide how to start. Should you start with the audio such as narration or start with the video? The most important thing is to just start, put something down on the timeline. Whatever motivates you, start there. But as you get more experience editing, you will probably notice that video is much

Figure 5.15 Production still: *Ephemera* (Chiacchio 2019).

more easily manipulated for timing than audio. If you are editing anything with a script, storyline or voice-over, then try to get your audio timed out first. Ignore the black holes and jump cuts you might have in the picture in the beginning and get your sound roughly timed out. Then go back and fix the video, replace video, put B reel on and adjust the timing of the audio as necessary. Don't get caught up for three hours colour correcting a shot that you may not even end up using. The story is the key and the audio is usually what drives 90 per cent of the story unless you are doing an artistic montage.

Another tip is to always have a metronome running in your mind and pay attention to it. We all have this sense of timing. It is not something you have to create. It's how we know when a joke is funny, when to break into a conversation or start a new sentence. Listen to your sense of timing as you are editing. If you're using background music, choose a song that doesn't conflict with the natural timing of your inner metronome. This is something you will just feel, not necessarily measure. If music is your main audio, that will likely influence the rate of your metronome instead of the other way around. Either way, paying attention to the timing doesn't mean you should cut on the beat. It is often much more impactful to put the action on the beat of the metronome and let the cuts fall invisibly between the beats. If someone is jumping, for instance, put their main burst of energy on the beat instead of cutting on the beat. Pay attention to the timing, respect it and know that you can translate your inner metronome to your timeline.

Most importantly, just go out there and get started, put your clips on the timeline and move them around, trim and start telling your story. It is amazing how many times fortuitous accidents happen when editing. Maybe you add music to some video on your timeline and just by chance the music hits at exactly the right place to make something funny or touching that wasn't before. Use it and build on it! Editing is like getting little surprises. You find them and preserve those things that work and get rid of the things that don't. It is like opening a bunch of gifts and because of that it can be very rewarding and fun.

We've been telling stories for thousands of years. And when Hollywood became the masters of telling a good story, it almost felt like you had to have some expensive gear, a crew, and know the right people to tell a good story. Now that's just not the case. The average person can tell a story that can impact the world and that's just amazing. My final advice is to be that person who gets up in the morning, shoots a story, edits it and shares it with the world.

Max: Fantastic, thank you.

5.3 Benoît Labourdette (France)

Max: Thank you very much for being part of the book, *Smartphone Filmmaking: Theory and Practice*, and we will dive straight into it. The first question: How did you get into smartphone and mobile filmmaking?

Benoît: I was always fond of little cameras. Before digital video, I was working with film and mainly super eight cameras and when in 1996 video appeared I had the first DV cameras. As a filmmaker, as an organizer of cinema events, festivals, etc., and as a teacher of making films I was always interested in mobile cameras. In 2004 the opportunity emerged in Paris to work with Mairie De Paris (City of Paris) and SFR (Société Française du Radiotelephone), a French telecommunications company. This initially quite commercial project asked me to imagine an event that we called Pocket Film Festival, an interesting event and I worked with CNC (The Centre national du Cinéma et de L'image animée) to drive the concept; this is not only publicity, this is not advertising, this has to have a real interest, and I found a great initiative. Why? Because I knew at the time, that ten years after, everyone in the world will have a camera in her/his mobile phone in one's pocket. And I imagined it will be something very significant and something that will change people's lives as the mobile phones have already changed a lot of things in our lives. We worked on the Pocket Film Festival for six years, with appropriate budgets and at the same time being able to engage in a very, very experimental approach, an event with quite lots of fans, filmmakers, audiences and makers.

Often when you make experimental things, you don't have fans and audiences, and this was invert and it completely astonishing for me to discover what people made with their mobile phone, and my spirit was all about discovering the potential. The idea was not to ask people to make a one-minute movie or this kind of competition. I am interested how we can invent a new cinema tool, as cinema was invented 123 years ago. At that time cinema was a new tool and people invented film languages, film forms and ways of using this tool. We discovered that a number of filmmakers were interested and even more engaging as not only filmmakers attended the festival. My aim was to each year make a different festival, listening and viewing mobile films and to be part of a community. And we discovered your work and it was one of the important works because you used the mobile phone in a specific way, a personal filmmaking style, and it would have been very difficult to make a feature film with a Nokia phone using 3pg video.

All these elements interested me back in 2004, and it still interests me today, because there are every day new tools, new apps, etc., which can continue the drive to explore new ways of filming. These days it is not as experimental anymore; it is the real world.

Max: And now mobile filmmaking is also an enterprise – but if we stay with the creative elements for a bit. Were there any mobile or smartphone films that stood out for you?

Benoît: I was always a filmmaker and an organizer of cinema events. And it is the same dynamic for me to show movies or making them myself.

In 2005 there is a short film, the title is *This Is Not a Movie*, in French *Ceci n'est pas un film* by Pascal Delé. It is a reference to the great surrealist painter René Magritte who made this famous painting *This Is Not a pipe* (also known as *The Treachery of Images*) in 1929. And this is a very interesting concept and one should remember that Pascal Delé filmed this short film with his mobile phone in 2005. He describes what we do today with mobile phones in terms of social media. He describes moving-images and picture as a conversation and not as a memory. For me this is an important movie because he imagined what happens now more than a decade ago. The short film is three minutes.

In 2006, *Mammah* was created by Louise Botkay-Courcier. *Mammah* is a film that is shot inside a hammam by a woman among other woman. And it is incredible because as a man it is impossible to go there. The mobile phone is not hidden. She films the girls and they look at the camera, so they know they are filmed that is accepted because it is filmed by a woman. The short film is part of the female body and it is part of the women's body language. The picture is so pixelated that there are connotations of *Peeping Tom* or voyeurism. A very beautiful movie and quite important to me.

In 2006 I made a feature film and it is a very extensive process to make feature films. The time between the shots, editing and showing it in public, this could be five years easily, and it is exhausting for the filmmaker.

Max: Yes, yes, yes.

Benoît: And that year, I decided, during each month to make one little film each day and put it on the internet, it was early beginning of YouTube at this time. It was completely new for me because I uploaded my short films to on the web. And the film *Série plans Séquence* (and after: *Les Acteurs Inconscients*) premiered on the web. By means of having the film online I had feedback from the audience immediately and I used the feedback to decide what I would film the next day. Then I combined these short films and it made a feature film (60 min). At that time this was completely new for me and I explored the possibility of these kind of experiences.

In 2005 the mobile feature *Nocturnes pour le Roi de Rome* (*Nocturne for King of Rome*) by Jean-Charles Fitoussi was produced. It was a feature film made by a young French director and it was an incredible work. I could not imagine that someone made a feature film with a mobile device and it was screened at the Cannes Film Festival, the year after in 2006, in the Semaine de la critique section.

In 2008 the mobile film *I Would Like to Share Spring with Someone* was produced by Joseph Morder. This was the first feature film shot with a mobile phone that was released in cinema theatres with a distribution deal.

Max: Of course, every project is different because you work with different people, you work in different locations, you work in different environments. But is there one element in your own films that stand out in the production process? How many films have you produced and created by the way?

Benoît: Me? With a mobile, smartphone or pocket camera?

Max: Yes.

Benoît: Lots of shots, perhaps fifty. I don't know exactly.

Max: Let's say fifty and that is probably conservative, within these films, is there something that you could think that is a common characteristic that appears in most of your work like a stylistic feature?

Benoît: One thing that is very important in my films and that interests me in this field is that the writing process has changed. Traditionally, you write a script and then you commence shooting. With these mobile cameras, it is possible to shoot and be inspired while shooting and even when producing fiction. Mobile cameras let us to be creative and to write movies directly.

I teach this technique to my students; the idea is to use reality, not to write a script and after that trying to turn it into reality. On the contrary, we go straight to reality and look and try, finding all the inspirations, all the possibility, all the places, all the actors, etc., in our environment and use that. As a consequence, one is not disappointed; i.e. I did not succeed in making what I thought I would create, I don't have enough money for this production, etc. NO, you use what you have and it is possible to make great things with what we have.

And if you make something in quite a classic way you will make a classic film. But I believe that audience are interested with new things as much as classic approaches. Furthermore, audiences enjoy strange or weird things. And in the Mobile Film Festival,

I experimented with the screenings that I organized, it is very important for me to mix different styles of mobile movies.

I am not fond of only one style of cinema and with these mobile camera phones, we are able to make all the styles of cinema. This can include professional production approaches with additional smartphone gear and crews, but it is also possible to experiment, and the two approaches can fuse. One can explore a classic narrative movie and also an experimental approach within one movie.

This mixing of genres and formats was always what I tried to programme in the Pocket Film Festival events. We can invent new ways of making movies. And for me now, we can continue to invent.

In this context we can also point at cinema in the theatre. About ten years ago Pippo Delbono made a feature film, *La Paura*, with my help, which was shot with his mobile phone. And it was released in cinemas and also featured in his theatre shows. He shot lots of videos himself that goes into the theatre shows. Before the time of mobile phones, he used video that was produced by other filmmakers. And now with smartphone filmmaking he can realize his visions. His mobile and smartphone filmmaking is outside the cinema context, but it shows a great contribution of mobile and smartphone filmmaking in the world of theatre.

Max: Do you think that one can work with a smartphone in a way that is difficult to produce with standard HD video cameras? Can the small size limitations become an advantage?

Benoît: Yes. This is an important question. For me, the specificity of the smartphone is that you have it every time in your pocket so you can take pictures and videos anytime. You do not have a photo camera on you every time, but you have mobile phone/smartphone always on you, we are a bionic human.

In the last years I worked a lot with drones (Figure 5.16). And the same image capture device and video technology that you find in a drone, you find in a smartphone. In the same way pocket cameras and smartphone are becoming one, i.e. the Sony cameras, the Eric 100 Sony camera that is a famous pocket camera and this lens can be plugged onto any smartphone. Fifteen years ago there was quite a difference between mobile phones and small digital photography cameras, but for me these tools are quite the same now.

A few years ago Black Magic Design released a pocket cinema camera; it is a very small camera with 4K video and the recent update has 6K video. For feature film productions this could be useful for action scenes or similar scenes. In a similar way the Canon 5D has been used for shooting entire feature films and similarly one can point of the potential of GoPro cameras, which were used in the feature film Leviathan, a documentary about a fishing boat including underwater cinematography.

Max: Taking about underwater cinematography. At the ninth MINA Screening the opening and closing films feature underwater cinematography. Have you done some work with apps, i.e. have you explored apps and post-production on tablets or smartphones or other innovative capacities of smartphones?

Benoît: In the last years I worked a lot with 360° video cameras (Figure 5.17). What I found interesting with tablets is the possibility to make the editing just after filming; it is possible too with mobile phones and smartphones, but the screens are quite small, and it is not always convenient. When I lead workshops on mobile, smartphone and pocket cinema, I often ask participants to make sequence shots or long takes. They are able to shoot, reshoot and shoot again. There is almost no more time between the shot and the moment

Figure 5.16 Pompidou Center (Paris–France): 'Drone altitude' workshop (Labourdette 2014).

Figure 5.17 La Fémis: Exploring Cinematic VR and 360° video (Labourdette 2017).

we see the movie together. And it is a very useful technique in a pedagogical way, because you learn a lot when you shoot something in sequence shots and long takes. In a sequence shot there is editing, but this is during the stage of the shooting (Figure 5.18).

Yes, this was possible before the advent of mobile and smartphones, but it is a great production approach for mobile, smartphones and pocket cameras. It is the intersection between your vision and reality. And the cinema for me, the sense of cinema is the ontology, is that this picture is the proof of what happened in reality.

And after that all the social media apps, like Snapchat, Instagram, TikTok, etc. Tick Tock is very interesting because the user makes edits; it is incredible what they make in this app. I do not use these apps, but in my workshop with young people, I use these apps because participants know these film languages. Some weeks ago, when I proposed to young people in a workshop to make movies, they said that is just like making TikTok movies.

Max: You mentioned Cinematic VR. In a similar way to the Sony lens, some VR cameras work with smartphone. And then of course, there is the whole world of AR happening. Is there something that really excites you? If you think back about the last ten to fifteen years of smartphone filmmaking, can you look forward to what could happen in the next five years?

Benoît: The new VR cameras have a lot of possibilities that we could not have imagined before. When you use a VR camera, there is a classic way to make 360° videos. But if

Figure 5.18 Port Zones Film Festival: Mobile films workshop and travelling projection (Labourdette 2016).

you look your film with the two spheres or in a VR mode that can be quite strange and quite incredible at the same time. And for me, what interests me, is to use these tools in new way and not always what they have been used for (Figure 5.19). I experiment and play with these tools, sometimes you respect the rules that the technicians gave to us and sometimes you have to break the rules and ignore technical standards (Figure 5.20). And here I can see a similarity between mobile, smartphone and mobile filmmaking, drones and Cinematic VR.

When drones arrived, it was quite the same. It was very difficult to control a drone. We think it is controllable, but it is quite uncontrollable. These cinematic machines have quite a bit of autonomy. The filming process can be challenging, due to the noise, flying the drones and the camera shake. But when you view the rushes and use the electronic stabilization, it is a completely different experience. Through exploring the drone videos one can discover a new world, a new vision of the world. This can be quite incredible. And with VR cameras, it is the same. These little machines have their own singularity. There is potential to open our minds to another way of thinking.

Now when I hold a camera, a regular camera or a smartphone, I hold the smartphone in my hand, and I am shooting as if I was using a drone.

Each new camera, for example, the Insta 360 ONE X, is a pocket VR camera. It has a stick, a very long stick and the stick is completely removed in the filming. You have a camera in space. As an artist, we are in very exciting times.

At this point in time art and cinema is a fusion between a machine and human. The machine is very important, so let's get started making our devices, similar to a painter who makes her/his own colours. Let's make our own machines or tweak the machines. And that is fun for not only the artists but also the audience (Figure 5.21).

Figure 5.19 La Fémis: Exploring Cinematic VR and 360° video (Labourdette 2017).

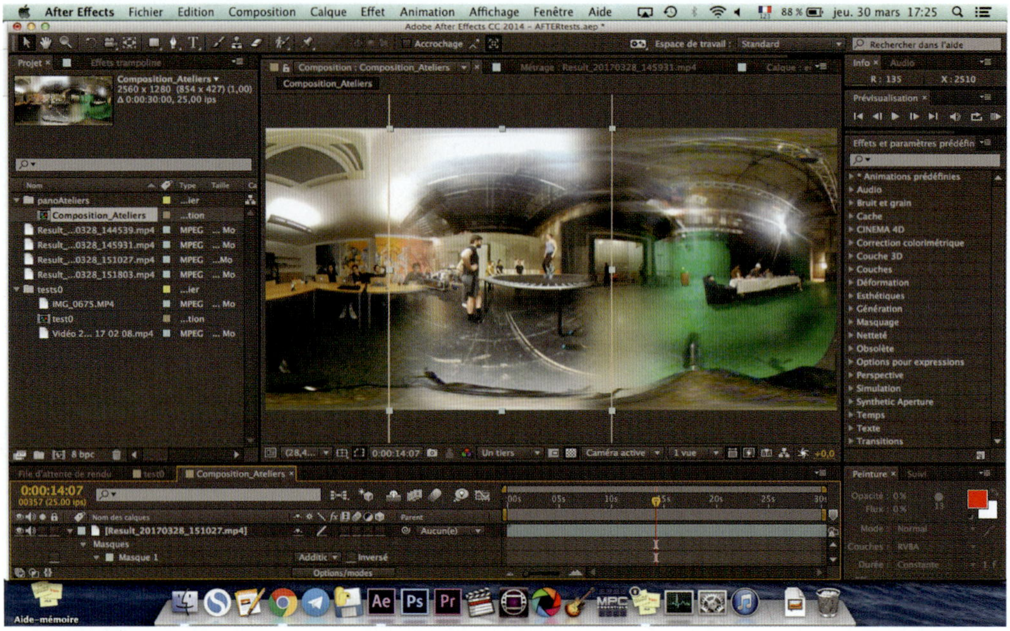

Figure 5.20 La Fémis: Exploring Cinematic VR and 360° video (Labourdette 2017).

Max: Do you think that smartphone film festivals are slightly different in some ways from sort of standard industry film festivals?

Benoît: When we made the Pocket Film Festival, I was questioning myself about the sense of this kind of event. Does it make sense? Because as you said, what is important in cinema is creativity. If the work is produced with a drone, VR or smartphone that is not the most important question, for it is what we create that counts. For me, there are perhaps two kinds of cinema, the film industry cinema and the art cinema.

The rules and conventions for an industry cinema, like the Cannes Film Festival, are very precise. Because the aim of all that is to send movies to cinema theatres, TV, etc., that is the aim. But even in this kind of industry film festival, you can find some different movies that stand out. Of course, the aim of these festivals is industry focused, and not experimental cinemas, and then there are other festivals, not only for experimental films but for films with different aims. Not mainly commercial aims but cultural aims.

These days mobile, smartphone and mobile films are shown at industry festivals and other places, and there is a place for everyone, and there is money for everyone too. This development came with the web sure. Here one could point at Chris Anderson, the director of Ted Conferences. In 2006 he released the book *Long Tail – Why the Future of Business Is Selling Less of More*. Within the context of mobile, smartphone and mobile filmmaking, that is quite important to understand. The idea of the mass market was before the internet – because now we can diversify and work in the niche market. Chris Anderson explained it very well. There are two economical models. There is the direct one, where you sell your film, and there is indirect one, because of your film you can do other things, etc., and make a living and sustain a creative career.

Max: As a smartphone filmmaker, educator and director of mobile film festivals, how do you engage with communities?

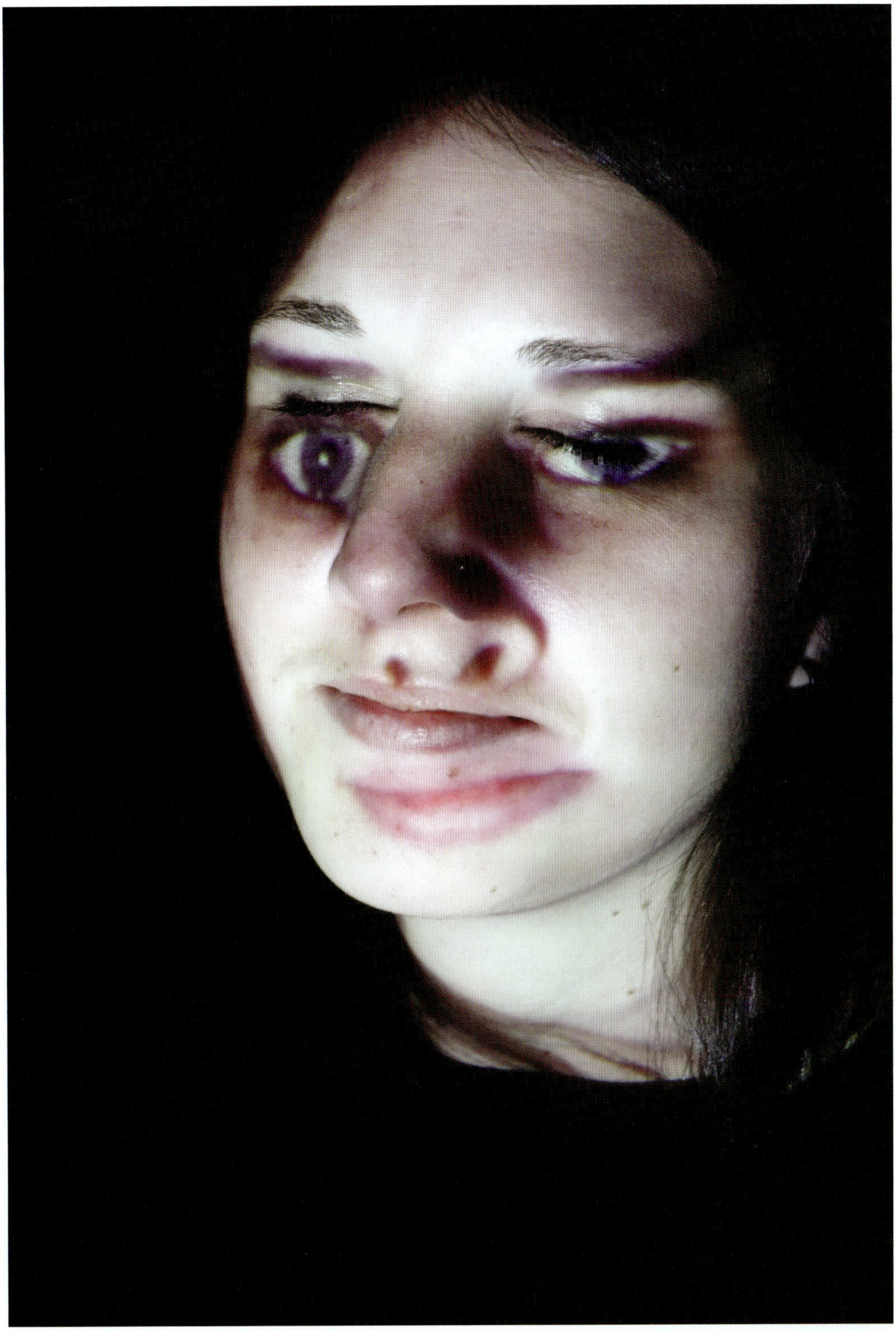

Figure 5.21 *Face on Face*, University of Grenoble (Labourdette 2017).

Benoît: For me, to be a filmmaker, to be an organizer of film events and to be a teacher is quite the same thing. In cinema schools, i.e. La Fémis, which is a renowned cinema school in France, I teach to people who will be the filmmakers in the industry. But I introduce them other ways of making movies, and I envision it should help them to be more free and more creative with their own creation. And I also teach adults in their professional development who are updating their knowledge. I help them to make serious films, with this non-serious mobile phone of their everyday life. And it seems obvious for me, for you, that it's possible to make something interesting with a mobile phone. In addition, I work with young people. I teach young and emerging filmmakers how to make movies. They learn about the language of the pictures, as they have to learn the language of the words. And in France I provide consultancy to governments and institutions, such as CNC, Town of Les Lialas, City Council of Nantes, Museum of Man, Ile de France Region. And I write books *Make a Film with Your Mobile Phone* (2007) and *Image Education 2.0* (2015) about mobile, pocket and smartphone filmmaking and articles on how to teach pocket cinema to young people, because those young people use their mobile phone every day. So, it is important for them to have a distance, a critical point of view, about their own practice (Figure 5.22).

I am working with psychologists about violence in suburbs, for example, and how we can make training sessions using mobile filmmaking for the people who are the social workers on these suburbs – to use smartphone filmmaking to contribute, to prevent violence, because the pictures are part of the violence issues.

Creativity, what we talked about, has to be mobilized at every stage. This creativity with the mobile, smartphone and mobile filmmaking is central our lives in the twenty-first century (Figure 5.23). And this also includes social power.

Max: Fantastic. Merci.

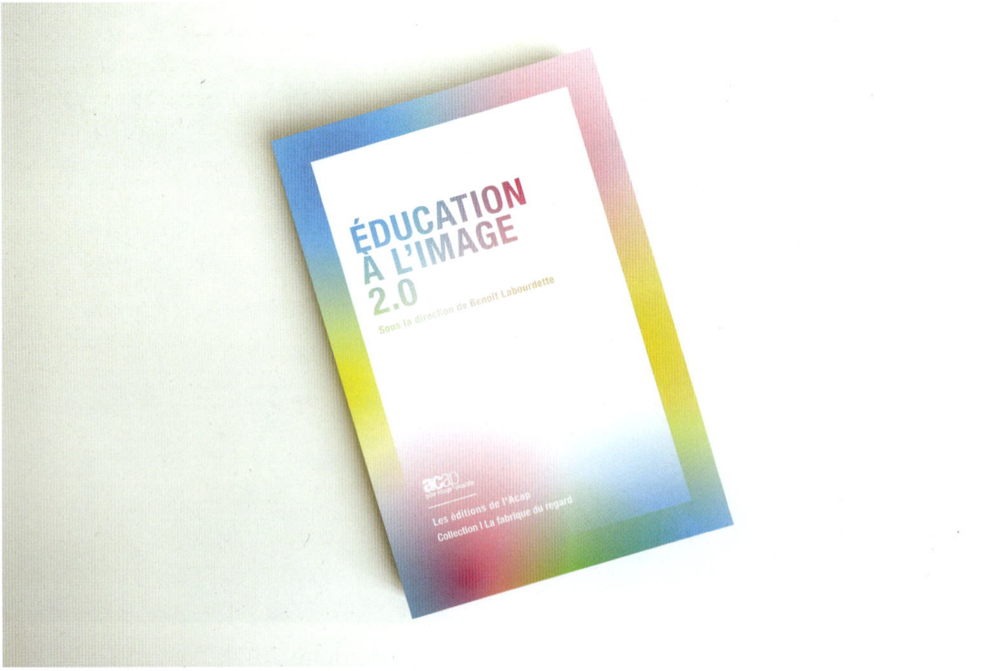

Figure 5.22 *Éducation à l'image 2.0* (English translation: *Image Education 2.0,* a new paradigm for the pedagogy of image) (Labourdette 2019).

Figure 5.23 Aubervilliers and La Courneuve (France): Films are on the walls! (Labourdette 2015).

5.4 Adobe Rush (United States and Asia Pacific)

Max: How did Adobe get into the smartphone filmmaking industry?

Adobe: Online video creation is bigger than ever before. Thanks to the rise of social platforms like YouTube, Instagram and Facebook, consumers want and need a steady stream of video content, available at the tap of a button. To meet this demand, a new class of online content creators have emerged, requiring tools that are easy to use and deliver professional-looking videos fast, right to the fingertips of their audience. Drawing on its long history of creating products for professional editors and filmmakers, Adobe worked closely with online video creators to build Premiere Rush, the first all-in-one cross-device video editing tool specifically designed for their needs (Figure 5.24).

In October 2014, what would go on to be the Adobe's first foray into smartphone filmmaking, Adobe Premiere Clip was launched as a free app on iOS. By this time, we had already recognized the growing need for burgeoning content creators to have a simplified editing experience, so we created a solution for anyone interested in making videos and not just professionals. Adobe Premiere Clip was launched with the ability to shoot, edit and publish from your phone in a time when there were not many options out there. We released the Android version in December 2015.

The learnings from Adobe Premiere Clip customers showed that there was not only a strong growth of smartphone video creators, but also an appetite to create more complex videos. This required further functionality and was an opportunity to build a deeper integration with the same artificial intelligence (AI) machine learning engine, which we call Adobe Sensei, that Premiere Pro takes advantage of, especially in audio with our Audition product.

As a result of our deep research and understanding of the evolving needs of this growing segment of the content creator population, we launched Premiere Rush in

Figure 5.24 Video producer on location (Adobe 2019).

October 2018 as a cloud-first application that can work between iOS, Mac OS and Windows 10 devices using cloud synchronization. This expanded the way content creators can start in one platform and continue work on the other. The Android release came in May 2019 (Figure 5.25).

Premiere Rush integrates intuitive editing, simplified colour correction, AI-powered audio clean-up, customizable Motion Graphics templates and seamless publishing to social platforms into a consistent user experience on desktop and mobile, enhancing its cross-device functionality. Understanding the importance of being able to work from anywhere and any device, Premiere Rush syncs all projects and media to the cloud so video creators can start editing on mobile and publish from their computer without ever missing a beat. And, for those who want to go deeper in their editing, Premiere Rush is integrated with Premiere Pro, allowing users to bring their projects to completion using Adobe's powerful editing platform.

With its similar interface to Premiere Pro, Premiere Rush is built upon the best user-centred insights around design making; it a seamless experience across multiple levels of expertise and use cases (Figure 5.26).

Max: Considering what happened in the last five or ten years what developments are you excited about?

Adobe: Smartphone developments in the last five years have seen an exponential increase in focus capabilities. Camera lenses with modern popular flagship phones now have three lenses providing a broad range of optical telephoto, standard and ultrawide abilities.

Smartphones are now competing with the DSLR and mirrorless cameras with software-based depth of field for video. The quality continues to pique with each generation of smartphone. Case in point, 4K is now standardized and device stability continues to extend to higher video resolutions.

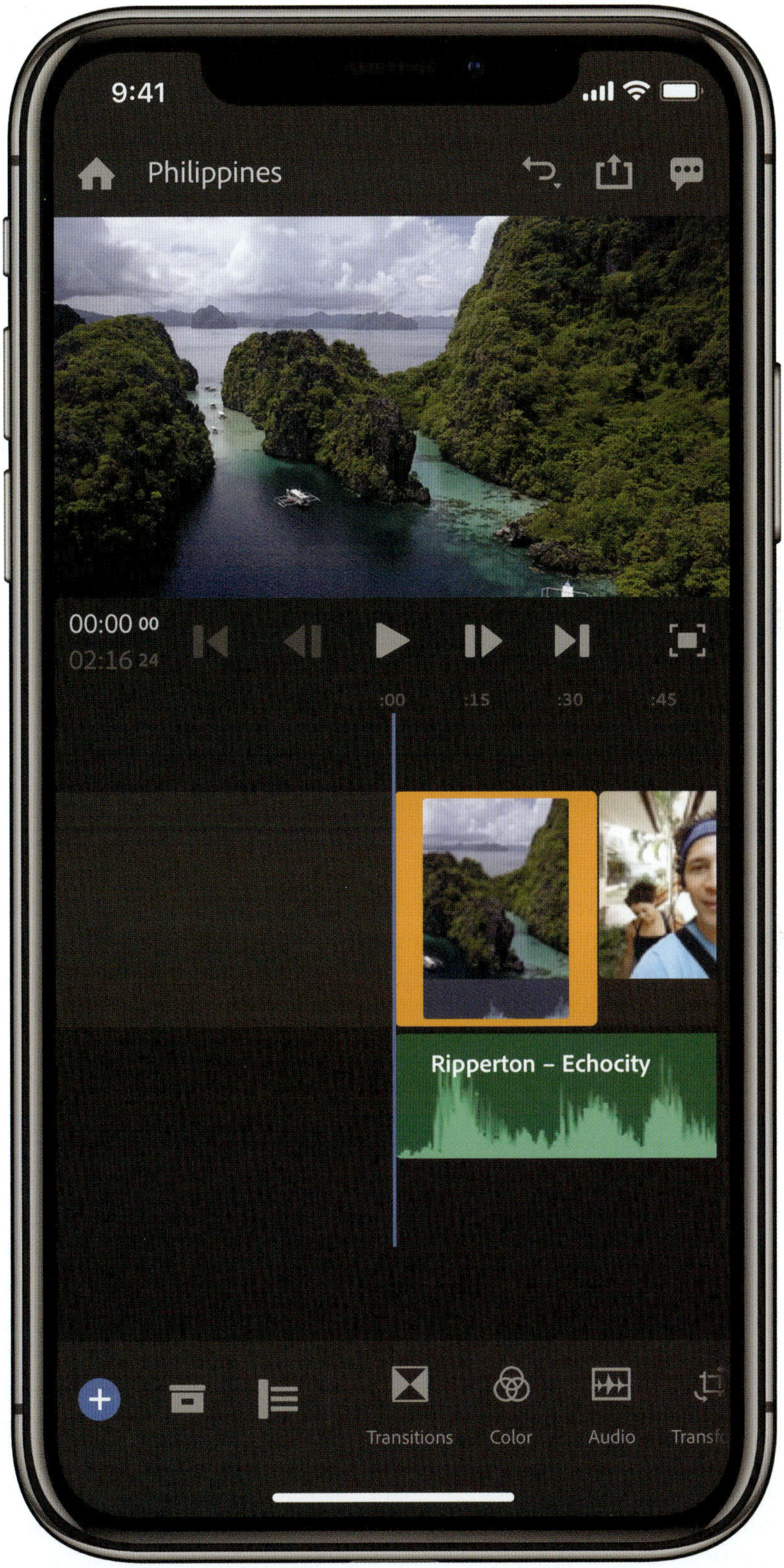

Figure 5.25 Adobe Rush interface (Adobe 2019).

Figure 5.26 Adobe Rush works on almost any devices (Adobe 2019).

Video frame ratios are a hot topic and Adobe has developed a feature in Premiere Pro 2020 that uses Adobe Sensei, Adobe's machine learning AI to auto reframe content for a chosen aspect ratio.

The most exciting developments in content acquisition, creation and output are clearly coming from the field of machine learning, which includes the current development cycles of stability, speech to text and auto framing for alternate ratios, auto ducking of audio under dialog and as the list of slow and repetitive tasks grow as will the machine learning solutions.

Max: How do you engage with the community of smartphone filmmaking and how do you see this developing?

Adobe: Smartphone filmmaking has continued to evolve, and the surrounding technologies have been developing very quickly with such hardware add-ons like microphones, gimbal stabilizers and a variety of software across capturing and editing.

Adobe engages with the video community through multiple online spaces and at events like VidCon and through our relationships with social media content creators who we work with regularly to fine-tune our product offerings and roadmap.

The smartphone filmmaking community is part of the broader filmmaking community as the same layers of story and developing craft apply across the spectrum of tools used in telling video stories.

Premiere Rush has helped not only online creators, but also media companies and creative agencies streamline their workflows and create content in a whole new way. We'll continue to see this trend develop as timelines, budgets and creative expression continue to evolve to satiate the general consumer's appetite for online and social media video content (Figure 5.27).

For example, VICE's UK office participated in a Premiere Rush learning event called a 'Creative Jam' in which VICE's team used the tool to capture footage in the field, piece their story together and publish it straight to social platforms. According to Tim Bertioli, VP, International Operations at VICE, 'The idea of an application that can make us faster to market and make more content is a game changer for us.'

Additionally, Premiere Rush's cross-device editing capabilities proved to be a boon for collaborative workflow and helped Havas Paris creative teams dissolve cross-functional

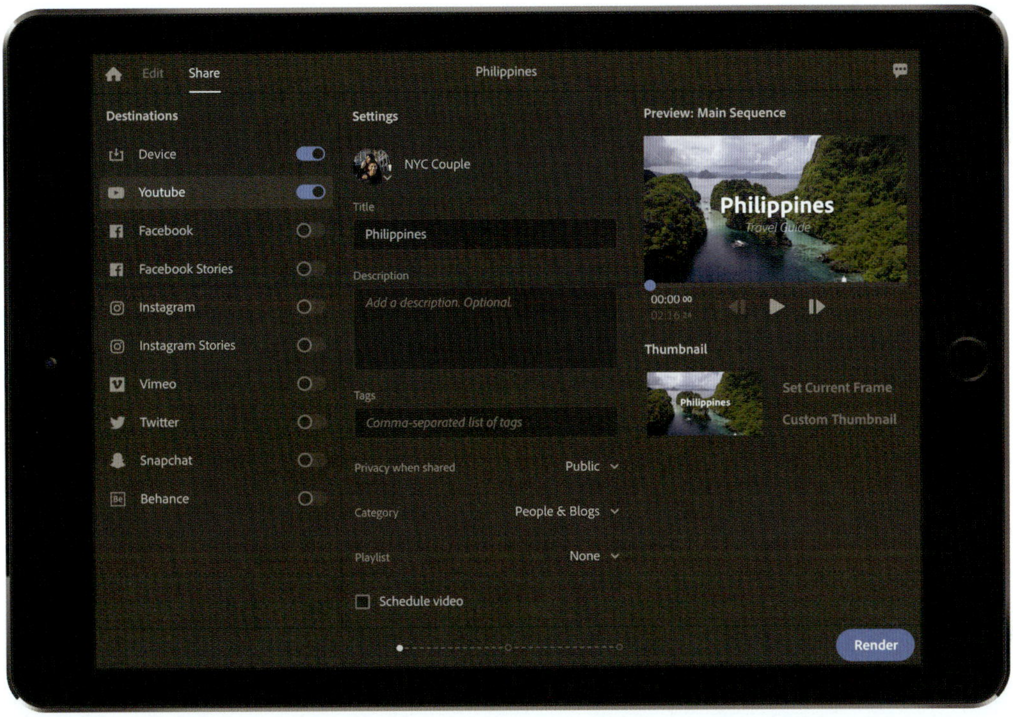

Figure 5.27 Adobe Rush. From script to screen to shoot and share (Adobe 2019).

barriers. 'Our work with Adobe was crucial to bringing creative staff together on a shared platform', says Thierry Grouleaud, Deputy General Manager, Havas Paris. 'The technology allows us to break down barriers between various creative professions.'

5.5 Treehouse (Australia)

Max: How did you get into the smartphone filmmaking industry?

Jason: My entry point was equally accidental as it was engineered. I had enjoyed a successful career in visual marketing and had written and directed numerous television commercials and short films. My turning point, however, was 2008, when the rudimentary mobile phone I owned at that time (initially a Sony Walkman W800 smartphone, then a Nokia N95) intrigued me with the capacity to capture basic video, not just digital stills. I well remember being curious about where this palm-sized technology would take us creatively and in fact challenged myself to create a short film on my Nokia N95 with a view to being one of the first filmmakers globally to try crack an international short film festival with a mobile phone-shot film (Figure 5.28).

The film I made (together with friends Shane Emmett and composer John Roy) was called *Mankind Is No Island*, and it premiered at Tropfest NY as the first mobile phone finalist to ever make it through to the world's largest short film festival. We won Best Film as well as People's Choice that night, and it sparked a string of festival awards and screenings, quickly gathering over 1,000,000 views on YouTube as well. It was also

Figure 5.28 Film still: *Mankind Is No Island* (Van Genderen 2008).

nominated for an Inside Film (IF) Award in 2009 (Best Short Documentary) which it won during a live telecast on SBS – to this day the only mobile or smartphone film to ever be nominated, let alone awarded, in IF Awards history.

Since then, my smartphone films have been awarded in dozens of festivals including Sundance London and Palm Springs, to name a few. Now, I run Australia's first smartphone storytelling studio, employing five staff, all working full time on creating everything from broadcast commercials to digital stories (Figure 5.29). I'm still pinching myself.

Max: Considering what happened in the last five or ten years, what developments are you excited about?

Jason: Computational imaging is quite possibly one of the largest progressive leaps that will exponentially grow smartphone cinematography. The App space is equally as exciting, working closely in partnership with leading smartphone brands to take advantage of quantum leaps in micro lens and chipset engineering.

Light field technology reimagines the physical depth of light our tiny lenses can capture and what we can now have increasingly finite control of. Focus will end up being something we can fix or change in post-production (like the photography app 'Focos' currently enables you to do to with portrait-mode stills). Backgrounds will be interchangeable. Lighting will even be completely correctable as software engineers build greater control over pixel maps and behaviours.

Folded Light Optics promises to pack expanded focal lengths into flat smartphones, running micro lens barrels flat inside our phone cases as opposed to on top of them. The Light L16 camera is a great early prototype of that technology, but we also know Apple

Figure 5.29 Jason Van Genderen's production set-up with Beastgrip and Beastgrip DOF Adapter MK2 model with Lonography Petzval 58-mm art lens for commercial productions (Van Genderen 2019).

and Samsung have both trademarked their own versions of that very same tech years ago. It is all in the pipeline. And that pipeline will be a complete industry game changer.

Personally, I would like to see imaging sensor engineers create a larger square or circular imaging chip, so we will no longer be tied to deciding if we want to shoot in landscape or portrait aspect ratio. That will end up becoming changeable in post-production too.

Max: How do you engage with the community of smartphone filmmaking and how do you see this developing?

Jason: I try to be as available for discussion and knowledge-sharing as I can be. I frequently present talks and workshops at film festivals all over the world, and am the founder of the Facebook group @FILMBREAKER, which was a community we created to help empower the smartphone creative community.

Most of my invitations to run talks are not to smartphone-save audiences; however, they are usually smartphone cynics! My conversations tend to be a little cyclic . . . continually exposing the creative power in our pockets and coaxing disbelievers into a new space to try.

In large, the film industry still refuses to believe that smartphones have any place even being on the starting blocks. Reimagining their expensive, gear-heavy industry seems too large a stretch for even fresh film school graduates . . . most still taking the point of view that serious cameras can only come from serious camera brands. I think RED proved that theory wrong with their failed Hydrogen One project . . . but not before numerous filmmakers parted with payments for the ill-thought out technology.

For me, the conversation still needs to be much louder and more frequent, and we need to get the new generation of smartphone cameras into the hands of doubters to convince them of the new possibilities. It is time we stopped trying to recreate what old cameras did with the new technology and time we started writing a forward-facing legacy for cinematography that will help give our Creative Arts industry jobs and opportunities.

Max: Are there any community champions that you work with? Can you let us know about their work?

Jason: Champions in this space unfortunately are still seemingly thin on-the-ground (Figure 5.30). But the people I admire and follow include Francesca Jager (@GetAheadMedia), new filmmaker Adrian Jeffs (Figure 5.31) and #MOJO champion Yusuf Omar (@Hashtag OurStories) (Figure 5.32). Simon Horrocks is a thought leader in this space too – and a generous giver of knowledge!

I think we are seeing some brave pioneers in the festival and exhibition space though, including Angela Blake and Alice Crew from (@SF3) SmartFone Flick Fest and Simon Horrocks and Andrea Holle from MoMo Film Festival.

Max: Can you share any best practice examples?

Jason: I can share some general 'best practice' thinking which would form a base to the way I teach smartphone cinematographers their craft. First is to know your tools and thrash them to breaking point. You need to make every mistake imaginable to fully comprehend the boundaries of the camera technology in your pocket. Shoot something in lousy low light with horrible grain so you can appreciate what control over your ISO will fix. Shoot everything auto so you know the problems that manual controls will overcome. Play with shutter speeds and see the difference they create, do not just read about them in a blog. You need to make the mistakes to own the solutions! Second big tip I would give is being intentional when you shoot with a smartphone. Don't film with it 'just because it was the only camera available'. Shoot with it because of its purpose and strengths. Choose it, don't

Figure 5.30 Francesca Jager (@GetAheadMedia) providing smartphone video training for industries, solo biz owners and entrepreneurs in New Zealand/Aotearoa (Jager 2019).

Figure 5.31 Still image: *Life Disrupted* (Jeffs 2018).

Figure 5.32 Yusuf Omar (@HashtagOurStories) is co-founder of Hashtag Our Stories. He is a correspondent working with wearable cameras (since 2010) (Omar 2019).

Figure 5.33 *Lost in the Crowd*. Jason with his co-producers Shane Emmett and Josephine Emmett (Van Genderen 2019).

let it choose you. Lastly, smartphone filmmaking demands a very different workflow and set of disciplines. Learn to master your own workflow that suits your production style and build on it.

Max: Is there anything that you would like to mention?

Jason: At the time of this interview, I am on the cusp of premiering my first feature-length documentary entirely shot on iPhone. It's called *Lost in the Crowd* and has been four years in production, shot almost entirely on iPhone 6s and 7. It looks at the darker side of crowdfunding. Even though I've been working in smartphone productions for what seems an eternity . . . this is a hugely weighty experience for me because it's my first step into longer format filmmaking. It also excites me because I know I will learn so much from how audiences and festivals will respond to this documentary . . . or not! Watch this space (Figure 5.33).

5.6 Struman Optics (Australia)

Max: Thank you for being part of the book *Smartphone Filmmaking: Theory and Practice*. It is fantastic to also have a passionate entrepreneur being part of this book project.

How did you get into the smartphone filmmaking industry?

Aron: It was funny when I first started using camera was an iPhone 3 and, I mean, originally it was an excuse for a camera. It was pretty basic, but I saw the potential of what it can do as things started improving. We purchased some lenses for the iPhone 4. The technique was okay, but it was more sold as a play toy than as a mainstream product or an accessory for a camera.

The quality of the smartphone was increasing in leaps and bounds, and I know there was going to be a need to accessorize a smartphone with some quality lenses even though originally it was for a photography. And we collaborated with the German engineering firm to develop some high-end smartphone lenses, something which photographers would even want to try it.

We wanted to bridge the gap between DSLRs and smartphones, at least in the beginning and bring smartphone filmmaking close to what DSLR cameras could do. We were approached by journalists, media agencies, filmmakers who started using our lenses and they said how great the video was (Figure 5.34). We then redeveloped the lenses to suit the latest camera smartphones. And our latest brand, we've just launched, YouShooter,[6] which we are proud to have co-branded with Rode Microphones, another global leader in sound technology and microphones (Figure 5.35). And now we have got Struman lenses,[7] which are being used by award-winning smartphone filmmakers, either short film or feature film, and also with the media agencies like Channel Nine, Channel Seven and ABC.

Originally, when we first started Struman, we went to a lot of marketing companies, and we were looking at the avatar of who our customer is going to be. And they said, 'Ah, it's going to be the tech savvy millennial person.' But it's been very, very different, and so we've had people from all ages, even kids using it. I've seen some incredible talent come out of young children all the way to tradespeople using it for taking photos and videos of their work. Parents are taking photos and videos of their children and sports. Obviously,

Figure 5.34 Struman Optics lenses (Struman Optics 2019).

music videos, young people creating a lot of music videos. And also, what's been surprising is . . . . It's surprising to me, because that's against everything all the marketing agencies originally predicted going back eight and nine years ago. Baby boomers, they use them to create travel videos, blogs, and a lot of people have been doing testimonials about products.

Max: This means the audience and 'user group' in smartphone filmmaking is potentially everyone . . .

Aron: Everyone now so . . . . Smartphone has given birth to a whole generation of videos, which we use every day.

Max: Considering what happened in the last five or ten years, what developments are you most excited about?

Aron: Just the fact real talent now can come out using smartphone filmmaking is really exciting. Numerous filmmakers, who were really restricted by the heavy cost of traditional camera gear and lack of opportunities or access to camera gear, are now with their own smartphones and accessories like Struman lenses, high-quality lights, microphones and can actually come out with incredible, talented short films and feature films (Figure 5.36).

And platforms like MINA and SF3 can put forward the creative work. Also, the great development has been with journalists, with news and the growth of digital media, reporters and journalists now are able to transmit live feeds without the heavy, cumbersome equipment and transporting. Smartphones are fast, discreet and safe.

Max: Great. And how do you engage with the community of smartphone filmmakers, and how do you see this developing?

Aron: We sponsor a number of smartphone filmmaking festivals and work closely with organizations like SF3, MINA, Filmbreaker Movement. We also sponsor a lot of equipment for schools in disadvantaged communities, where they engage in creating short films, hence putting forward powerful messages in the form of short films and giving them a voice.

Figure 5.35 Struman Optics YouShooter (Struman Optics 2020).

Max: Fantastic. That's really great. And are there any community champions out there that you work with?

Aron: Yeah, we work with . . . . They are all champions in their own way. Every single person who has put out a movie, it is got a powerful message. Like they say, we have got 6 billion stories and counting.

Rob Morrison was chief reporter of Channel 9. He is one of the important figures because he is extremely passionate about smartphone journalism in the way of storytelling and reporting stories. And he is also an integral part of YouShooter, which is a video lens brand. He has won several awards for journalism, including reporting breaking news, live, with the help of a smartphone.

And Adrian Jeffs, who is an award-winning short filmmaker, who's shot one of his award-winning films entirely with our lens.

Max: Are there any other best practice examples that you can share?

Figure 5.36 Struman Optics lens on location (Struman Optics 2019).

Aron: Absolutely. For businesses especially now, smartphone is an integral tool, not just about communicating and talking on the phone. It is about explaining their brand and putting forward their story. No longer is it viable just to put an ad on yellow pages and sit there and wait for the customers. Social media has become an extremely important part of what a business does and how they communicate with their customers – either through Instagram, Facebook or new emerging platforms, and they communicate through Instagram Stories, Facebook Stories and live feeds, and making every day videos. Corporations ranging from blue-chip to small enterprises are using video; the human resources department is making training videos in the form of stories . . . even medical professionals or doctors now, I know are using mobile phones to record things, to send it to their colleagues, get second opinions. So, there is a lot happening in the space of mobile phone, smartphone, video and storytelling (Figure 5.37).

Max: Are there any projects or people that inspire you?

Aron: Definitely, there are lots of people who inspire in this industry. Every day I am surprised what can come out of this small mobile phone. Recently, we sent some lenses to an Aboriginal community in Northern territory, and they had some really powerful messages where they wanted to communicate with the Cloncurry Shire Council and they actually made it in the movie and put their messages forward. That's the kind of strong response you are getting with storytelling rather than just writing a letter, send them videos.

We recently had one of the SF3 winners, it was a sixteen-year-old 'youngstar' who competed and won. And for someone to have that kind of talent among doing every other thing a schoolchild would, it's just inspiring.

Max: Can I ask you one more question about the journey into smartphone filmmaking . . . . Before you started making the lenses, did you think about the film industry, or were you working in a different sector?

Figure 5.37 Struman Optics lens in action for enterprise video production (Struman Optics 2019).

Aron: Growing up, I had a few school friends whose parents were into filmmaking. As a child, I used to watch this and there used to be a big crew, massive cameras, people holding lights. This whole scene of video making and filmmaking was very different. Later on, in my life, I saw people just with a mobile phone, being able to create something and the death of the home video cam, we all had uncles when we were young, someone who used to make home videos. And I cannot remember the last time I saw someone with a video camera. That thing has completely been replaced. Our mobile phones have replaced alarm clocks, pagers and torch light, and everything. It is just a natural progression to replace a camera. And I was excited to be part of that, to be able to contribute, and I was inspired that I could create a product which could make a difference in that field.

Max: Fantastic. Thanks. Is there anything that you would like to mention that I have forgotten?

Aron: Struman, as a company, we are inspired to create products not just for the mass market. The problem with mass market products is they have to be cheap and something which can be cost effective to bring it to you. We wanted to create a product which would stand the scrutiny of an avid user, so someone who is an experienced user and still be cost effective. We are not the biggest company manufacturing lenses; we definitely want to be the best lens in the market. When we actually manufacture lenses, even the costs are small, because we do not manufacture millions of lenses, so a few dollars more for each lens is worth for us investing because that is our point of difference (Figure 5.38).

My passion really is to create a programme for school kids and disadvantaged kids to utilize their most used device. Every child, I know quite often they get blamed for staying on it too long. To give them the tools to capture either photography or videos not only as an art form but also as a medium to express their feelings, communicate their concerns,

Figure 5.38 Struman Optics lens: Cinematic wide (Struman Optics 2019).

keep them occupied, kill boredom, reduce anxiety and ultimately, depression. If we can save one life by reducing suicide as an ultimate goal, that'll be time worthwhile spent. And we are passionate to be working on that.

Max: Fantastic. It's really great. Let's work on this. Thanks for being part of this book.

Aron: Thank you.

Notes

1. Rencontres de la vidéo mobile, https://video-mobile.org.
2. SmartFone Flick Fest, https://sf3.com.au/.
3. SmartPhilm Fest, http://www.smartphilm.com/.
4. MINA Mobile Innovation Network and Association, http://www.mina.pro/.
5. *Ephemera* documentary, http://www.ephemera.life.
6. YouShooter, http://youshooter.com.au/.
7. Struman Optics, https://strumanoptics.com.au/.

References

Anderson, Chris. 2006. *The Long Tail – Why the Future of Business Is Selling Less of More*. New York City: Hyperion.

Labourdette, Benoît. 2007. *Tournez un film avec votre téléphone portable* (Make a Film with Your Mobile Phone). Paris: Dixit.

Labourdette, Benoît. 2015. *Éducation à l'image 2.0* (Image Education 2.0). Amiens: ACAP.

Films

Ephemera. Dir. Angelo Chiacchio. International: independent production.

J'aimerais partager le printemps avec quelqu'un. 2008. (I Would Like to Share Spring with Someone) Dir. Joseph Morder. France: Forum des Images.

La Paura. 2009. Dir. Pippo Delbono. France: Quidam Production, Les Films d'Ici and Forum des Images.

Les Acteurs Inconscients. 2008. Dir. Benoît Labourdette. France: independent production.

Leviathan. 2013. Dir. Verena Paravel and Lucien Castaing-Taylor. United States: Cinema Guild).

Lost in the Crowd. 2019. Dir. Jason Van Genderen. Australia: Treehouse Creative.

Mammah. 2006. Dir. Louise Botkay-Courcier. France: independent production.

Mankind Is No Island. 2008. Dir. Jason Van Genderen. Australia and United States: Treehouse Creative.

Nocturnes pour le Roi de Rome (Nocturne for King of Rome). 2005. Dir. Jean-Charles Fitoussi. France: Aura Été.

Ceci n'est pas un film (This Is Not a Movie). 2005. Dir. Pascal Delé. France: independent production.

6

Creative innovation

Chapter Ouline

Sean Baker's *Tangerine* can be described as the first smartphone film to tour all major international film festivals including a premiere at Sundance Film Festival (Salt Lake City, United States, 2015) and then showcased with screenings at San Francisco, Seattle (United States) and Sydney (Australia) film festivals, among others. *Tangerine* was produced with a $100,000 budget using the iPhone 5 (Watercutter 2015). Baker used the anamorphic adapter from Moondog Labs[1] and the FiLMiC Pro app[2] using a Steadicam rig, which are standard production tools in the international smartphone filmmaking community. *Wired Magazine* reported that Baker used Instagram, Vine and SoundCloud to scout talent and music (Watercutter 2015). He applied smartphone filmmaking techniques that are reminiscent of street photography or the French New Wave. Using natural light, filming on location and following the actors through environments of everyday people, not always in the limelight, characterize *Tangerine* as much as smartphone films by the pioneers or La Nouvelle Vague. In Baker's feature, the narrative intertwines immigrant and transgender communities and speaks to the capacity of smartphone filmmaking getting as close to the subjects as possible. Following the success of *Tangerine*, Baker produced *Snowbird* (2016) for the fashion brand Kenzo.

The breakthrough for smartphone filmmaking in the Hollywood blockbuster world was Soderbergh's *Unsane*. A slightly unsettling and not aimed at perfection crispness of images is seen in *Unsane*. Steven Soderbergh used the iPhone 7 and had about a $1.5 million budget to make this American psychological horror (McClintock 2018). A closer look at *Tangerine* and *Unsane* on IMDb (Internet Movie Database) reveals that both Soderbergh and Baker personally engaged in the cinematography. On IMDb[3] Soderbergh is credited alongside Peter Andrews, using his artist name as DP (Director of Photography), which originates from his dad's first and middle names (Snyder 2007). On this note, one should mention that the editing credit Mary Ann Bernard is also Steven Soderbergh (Zoller and Carvajal 2013). Baker worked with the DP, camera operator and co-producer Radium Cheung.

Both projects are characterized by smaller crews where teamwork cuts across traditional fixed roles as known in studio productions. Film set crews were smaller intentionally allowing movement at any pace through the streets of Los Angeles or through interior environments such as hospital corridors. Both Soderbergh and Baker did not reduce wo/man-power in the sound departments, ensuring high-quality sound recording and design. While Baker works with seven mixers, dialogue editors, Foley artists and sound mixers, Soderbergh has a much larger sound department of twenty-one crew members. Similarly, Baker has nine editors working with him and Soderbergh, a team of three editors including a VFX department of five. With *High Flying Bird* Soderbergh also introduced smartphone filmmaking to Netflix in 2019. His films demonstrate the significance of independent filmmaking and the studios' search for creative innovation. Soderbergh developed a track record of producing Hollywood blockbusters such as *Out of Sight*, *Ocean's* (*film series*) or *The Informant!*, and as an independent director, he is equally respected by audiences and peers for his independent filmmaking style that characterizes his films such as *Sex, Lies, and Videotapes* or *Solaris*. As an independent filmmaker, he does not necessarily divide the production process and roles as traditionally set by the studio system. Soderbergh, as much as Baker, reveals the emergence of the director-cinematographer. In filmmaking, the writer-director role is well established, but it is notable that several smartphone filmmakers are director-cinematographers. This also applies to Horrocks and Van Genderen, who are featured in the previous chapters. It is no surprise that Astruc's vision is now realized in the independent director-cinematographer.

In 2019 Academy-Award-winning director Claude Lelouch premiered *The Best Years of a Life* at Festival de Cannes.[4] In a recent interview he reconfirms the notions that smartphone filmmaking can create novel experiences, points towards the significance of locations in filming and talks about the unique connection smartphone filmmaking can create with an audience:

> It's really, really good. It delivers a picture that can transmit an emotion. Really. Because it's so small and light you can put the camera everywhere, it can move fast. It's like a painter discovering a new colour. This is a new colour. If you have a shot where people are running, it's perfect for that kind of shot, for example. You can do whatever you want even in situations which are normally hard, like a confined space.
>
> (Phelan 2019)

This argument very much reflects the mobile video and moving-image characteristics as outlined in 'Aesthetics of Mobile Media Art' (Baker, Schleser and Molga 2009) and defined in the 'Keitai aesthetic' (Schleser 2011) about a decade later, confirming their significance in contemporary smartphone filmmaking. More important than the technical and optical standards that match broadcasting and blockbuster standards is the creative freedom to experiment with mobile, smartphone and pocket cameras; smartphone filmmaking provides a licence for creative experimentation. Lelouch worked with a cinematographer, Robert Alazraki, but also contributed to the filming producing scenes with an iPhone and a custom-made gimbal iPhone rig (Falt 2019). Next to the narrative dimension, the aspect of co-creation and engagement of audiences as much as its authenticity, resonate in smartphone filmmaking and define the development of this film form. The small lens and size of a smartphone allow filmmakers to capture subjects and locations in virtually any camera position. Composition and framing can work in the context of traditional cinematic techniques, as well as through new approaches such as vertical video or selfies (Figure 6.1).

At the 59th Berlinale two smartphone feature documentaries, *Midnight Traveler* by Hassan Fazili and *Selfie* by Agostino Ferrente, were screened in the Panorama Dokumente programme.[5] As the name indicates, the film by the Italian filmmaker Ferrente is filmed from a first-person perspective. Rather than documenting the stereotypes of two Neapolitans (sixteen-year-olds), he handed the smartphones over to the teenagers, Pietro Orlando and Alessandro Antonelli, to document their everyday life in the Traiano district of Naples. The story reveals the tragedy of their friend Davide Bifolco who was killed by the police. The tragedy happened as the police mistook him for a fleeing criminal. While the first few minutes might be quite different from other film and documentary experiences, and one might take time to adapt to the selfie camera position of the two teenagers, after 10 minutes one feels like one of their best friends. This smartphone documentary demonstrates the intimate characteristics as the film could only be realized with a smartphone video camera and the documentary subjects (and not the filmmaker) – the teenagers' film in public, in their homes and among their other friends. The result is an authentic account that documents a community from an insider perspective. One can compare the aesthetics more to a Facebook post than a BBC documentary, but its connection to the audience is key for this subject and story. It is free from preconceptions and provides an experience of everyday life for these Italian teenagers. It is with great

Figure 6.1 Film still: *Selfie*. Pietro Orlando and Alessandro Antonelli (Ferrente 2018).

Figure 6.2 Film still: *Midnight Traveler* (Oscilloscope Laboratories 2018).

interest to see that the broadcaster ARTE France supported the production, which was also broadcasted internationally following the festival circuit (Figure 6.2).

Filmmaker Hassan Fazili and his family documented their refugee journey from Afghanistan to Europe using their mobile and smartphones. They captured their everyday life, and his wife Fatima Hussaini and two daughters Nargis and Zahra (three and six years old) contributed to the production during their approximate three-year journey. The documentary film also reflects on storytelling and how the family dealt with their situation of extreme fear, hope and many dangerous situations crossing borders and facing with traffickers and racist attacks on asylum centres. *Midnight Traveler* provides insights to the Fazili family and other refugees on the run and in asylum centres. The mobile phone is constantly with the family, and their intimate account captures their perspectives. As we get to know the family their journey, this documentary resonates the early mobile media art aesthetics with its emphasis on location and recognizing the connectedness to the filmmakers. The film was shot by the entire family using three mobile phones, Hassan's, Fatima's and Nargis's mobile devices. Su Kim, an Academy Award–nominated producer, worked with Emelie Mahdavian as producer, writer and editor. Old Chilly Pictures received support from Fork Films Threshold Foundation, Sundance Institute Documentary Film Program, Doha Film Institute, Just Films, the Hot Doc Crosscurrents Doc Fund, Tribeca All Access, among others. Following the international film festival screenings, the documentary received theatrical distribution and was boradcasted internationally.

At the 70th Berlinale, *Saudi Runaway* premiered and received a distribution deal by National Geographic. Swiss-German filmmaker Susanne Regina Meures co-created the documentary film with a young woman named Muna, whom she met and contacted via an online messaging group chat. The Saudi Arabian twenty-something decided to flee the country on her honeymoon as this was the only time she was allowed to travel. Muna documented her life for about five weeks before and in the days leading up to her wedding

Figure 6.3 Film still: *Saudi Runaway* (Christian Frei Film Productions 2019).

as well as her escape shortly after the arranged marriage. The film reveals her desire for freedom to follow her aspirations, unobtainable within this totalitarian and patriarchal culture. She shares her life and courage with us, which Meures weaves into a documentary narrative that could not have been scripted with more suspense (Figure 6.3).

Next to the recognition at major festivals, major commissions include *Détour* by Michel Gondry via Apple (Figure 6.4). Michel Gondry is well known for *Eternal Sunshine of the Spotless Mind*, and he is the winner of the Academy Award for Best Original Screenplay, *Mood Indigo*, and has created music videos for The Chemical Brothers and Björk among many other creative and commercial film projects. As part of the *Détour* project, he also created some great behind-the-scenes videos featuring the iPhone's film production capacity and possibilities for time lapse, slow motion, stop motion, unconventional perspectives and night scenes. These videos as much as *Détour*, a witty story of a young girl's bike that is lost on a holiday trip, demonstrate smartphone filmmaking's creative capacity and potential for imaginative storytelling. In 2019, Apple's *#ShotoniPhone* commissioned Donghoon Jun and James Thornton of Incite to produce *Experiments*. The creative technology studio and production company from Los Angeles showed the detail the iPhone X can capture. *Experiments* is a celebration of organic shapes captured in '4K, slow-mo and time-lapses',[6] in a 'bullet time rig' for 'mind-bending 360-degree footage'[7] with a focus on the natural elements of water, fire and ice. This creative exploration to find immersive and imaginary worlds can be facilitated by the smartphone as an experimentation device. While the camera can not only be turned in any direction and the small lens is an advantage for the super close-up, a key characteristic of smartphone filmmaking, Incite also used multiple iPhones to build a bullet-time camera rig. Conceptually, the videos relate to the earlier discussed mobile moving-image work *Sketch-Three: Avant-Garde (R.P.M. 2)* (Figure 6.5) or *The Big City* .

Figure 6.4 Michel Gondry. Behind-the-scenes video for *Détour* (Gondry 2017).

Figure 6.5 Film still: *Sketch-Three: Avant-Garde (R.P.M. 2)* (Fox 2014).

6.1 Experimentation driving innovation

The previous sections discussed the innovation in terms of aesthetic refinement and development of an original film form and format specific to smartphone filmmaking. The next section will further explore the collaborative elements in the participatory and online culture. Now that almost anyone in the world has a mobile device or smartphone, could a film be produced collaboratively? How can creativity engage audiences and community groups? Rather than only focusing on the final outcome in the form of a smartphone film, this research also speaks to novel processes and innovative formats.

As a filmmaker, Schleser has produced more than a dozen mobile, smartphone and pocket short films and two feature-length documentaries. Through the application of practice-led research, his work explores the constant aesthetic refinement of smartphone filmmaking. The aesthetic explorations drive the conceptual engagement and understanding of mobile media, shifting from a visual plane to a point of collaboration. Through developing non-linear formats and exploring storytelling for social innovation, smartphone filmmaking as a film form is continuously redefined. It is critical to note that mobile, smartphone and mobile filmmaking contributes to art and culture beyond moving-image, film and video industries. The accessibility of smartphones affords novel approaches to collaboration and provides a means to engage audiences. Until very recently, no frameworks specific to smartphone filmmaking existed that allowed a focused analysis. While there is a vast amount of technical innovation in all areas of production, it is key to also consider social dimension to differentiate mobile media (based on peer-to-peer networks) from broadcast media (one to many distribution). In the 2012 article 'What Is the Future of Mobile' (Sterkenburg 2012) Schleser argues for mobile innovation as a model that aims to be sustainable and should enquire and strive to establish strong social interactions and links through engaging local communities.

> In this context one should not only consider mobile media as screen media like T.V. or desktop computers, but a social interactive and multi-nodal network media that links to communities, provides new insights through visual communication, while providing access in many respects for a global audience. The notions of sociability and connectivity are and will be key in the future of mobile media.
>
> (Schleser 2012 online)

These qualities of smartphone filmmaking, which cannot be replicated by DSLR, Blackmagic or RED cameras (and any other standard-size digital video camera), are key in collaborative storytelling. The early mobile media aesthetics established the intimate, immediate and referenced the significance of location as key signifiers of mobile and now smartphone filmmaking. Through linking these screen-based qualities into the domain of online and thus networked videos, one can argue for a development towards collaborative peer networks. This model is aligned with Anderson's *Long Tail* (2004) approach and recognizes the internet as a place for niche audiences. The political dimension is outlined *Smart [Phone] Filmmakers >> Smart [Political] Actions* in detail (Schleser 2018). Schleser reviews three smartphone film projects, *Reel Health* (Tanzania by Tigo, Remedee, Think/Feel and Touch Foundation), *Spirit of Rangatahi Mobile Filmmaking Workshop* and *In Response We Closed Flinders* that were screened at the MINA screenings. These smartphone films embrace open frameworks, including more collaborative and reflexive storytelling approaches (Schleser 2018, 121). In an academic context, these smartphone films and the ones outlined in the next section can be analysed through the *Open Space Documentary* framework. Writing in *Open Space New Media Documentary: A Toolkit for Theory and Practice* Zimmermann and De Michiel define collaborative projects through small places, designing encounters, polyphonic collaborations and inviting spaces (2017). The analysis presented in this book is aligned to the innovation in the filmmaking process:

Figure 6.6 Max Schleser making mobile films for *Pangea Day*. Filming and editing with other participants on Nokia N95 (Schleser 2008).

> Open Space documentary shifts away from a narrow focus on the highly crafted finished product toward responsive and iterative processes deployed across platforms and places.
>
> (De Michiel and Zimmermann 2017, 144)

Zimmermann and De Michiel define complexities as perspectives, experiences and viewpoints in dialogue with each other (2017, 106). Next to complexity, they identify the working principles of *Open Space Documentary* as Circularity, Collaboration, Community, Composting, Connection, Context, Continuum, Conversation, Cost. These Cs define the innovative elements of the collaborative projects *24 Frames 24 Hours* (Figure 6.7), *Viewfinders* and *#Nucleus* (Figure 6.8).

6.2 Transformational creativity

In the context of collaborative storytelling, one of the most pioneering mobile films was the *Pangea Day* project, which brought mobile phone filmmakers from around the world together in a celebration on 10 May 2008 during a four-hour live broadcast in Cairo, Kigali, London, Los Angeles, Mumbai and Rio de Janeiro. Nokia partnered with and sponsored documentary filmmaker and Ted Prize winner Jehane Noujaim, who wanted to connect

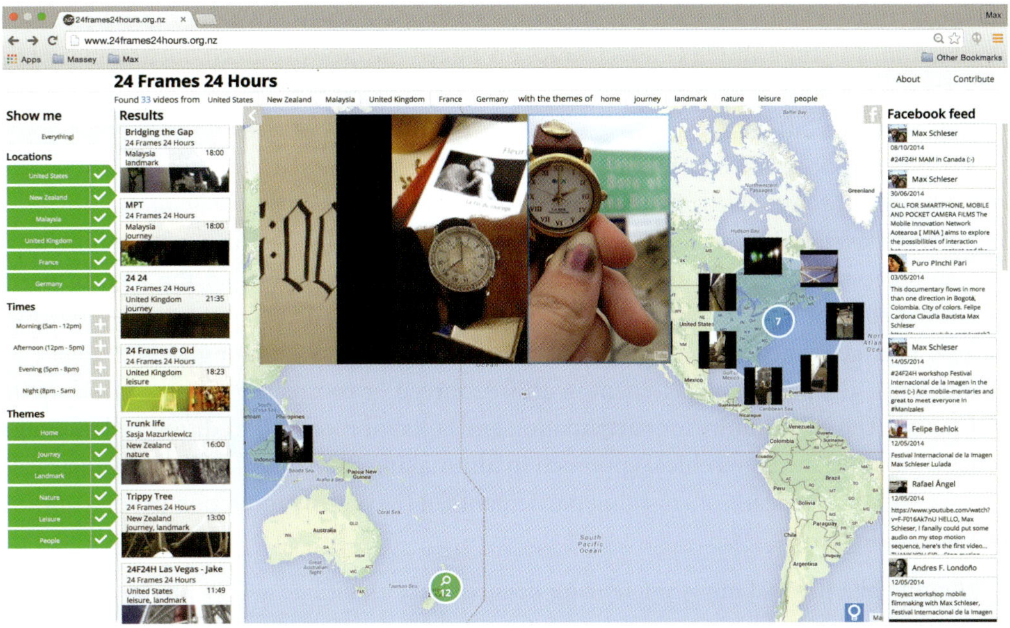

Figure 6.7 *24 Frames 24 Hours* interface for user/participant-based storytelling (Schleser 2014–17).

Figure 6.8 Production still: *#Nucleus* smartphone transmedia workshop (Goethe-Institut 2019).

the world through film. In a similar, but conceptually quite distinctive, approach Scott Free Production, YouTube and L. G. Electronics produced the crowdsourced documentary *Life in a Day* on 24 July 2010. The film premiered at Sundance Film Festival (Salt Lake City, United States) in 2011.

Within collaborative smartphone filmmaking, the process can be as important as the outcome of a project itself. In the projects *24 Frames 24 Hours*, *Viewfinders* and most recently *#Nucleus*, Schleser explored the concept of producing a smartphone film with peer-audiences (Cammaer and Schleser 2018) and further developed the idea of user-generated content towards a framework for workshop-generated content (Schleser 2018). Several user-generated content projects, such as *Life in a Day*, do not give any agency to participants. The level of engagement in crowdsourced or co-created projects can be quite distinctive. The opportunity for participation and collaboration in the editing and montage process can shift the hierarchical approach in filmmaking from a top-down model towards a conversation. By means of engaging participants in the entire production process affords an open dialogue:

> Here mobile filmmaking is referencing documentary traditions and illustrates how these concepts can be applied to social media. Contemporary crowdsourcing strategies not only question notions of authorship, but also the examination of these works in terms of their understanding of creativity. Collaborative mobile filmmaking shifts away from looking at film as an end product towards examining the processes that can be revealed. . . . The editing process can be understood as a negotiation and as an open dialogue rather than a linear construct. *24 Frames 24 Hours* creates a reflexive process that allows participants to engage in a subject of their choice that is relevant to them. Through the creative involvement, collaborative mobile filmmaking creates a sense of presence and sociability. . . . The self-reflexive process is an indicator for the realisation of audience engagement in collaborative mobile phone filmmaking.
>
> (Schleser 2012, 86)

Collaborative smartphone films embody people, places, and technology in order to 'join together to visualise different ways of acting in the world with others, and to imaging new sustainable futures' (De Michiel and Zimmermann 2017, 144). *#Nucleus* is a transmedia experience that celebrates and creates awareness about local nature and the environment through smartphone filmmaking. Ten filmmakers from seven Australasian countries were invited to a workshop at the Goethe-Institut (New Zealand). Max Schleser, Michael Hollis (Australia), Fauzhyana Sharifa and Nabil Azizi (Indonesia), Amirul Firdaus (Malaysia), Sarah Davis (New Zealand/Aotearoa), Loyd Doron (Philippines), Tejas Ewing (Singapore), Huy Tam Dang (Vietnam), Andrew Robb (Australia) and Christian Kahnt (New Zealand/Aotearoa) co-created an 11-minute eco-smartphone film. Also, aligned with these elements, Schleser created a 7-minute smartphone film responding to the research questions; what does ecotourism mean and what effects does travelling have on the environment? And how can smartphone filmmaking inspire new connections with people and places? (Figure 6.9) Both of these short films were presented in a public screening in Wellington's Embassy Cinema (New Zealand) and then selected for and screened at the Super 9 Film Festival in Portugal in 2019.

Figure 6.9 Production still: *#Nucleus* smartphone filmmakers on location (Goethe-Institut 2019).

Elements of indigenous storytelling inspired the above research project and through the Māori concept of kaitiakitanga created new affinities and connections with people and places. This concept denotes stewardship and care for the land and world around us, and in combination with smartphone filmmaking facilitated a democratic approach to storytelling exploring issues of the environment and sustainability. The practice-led research into innovative creative processes for transmedia productions resulted in the development of a creative method including community engagement through smartphone filmmaking workshops in New Zealand/Aotearoa (at Auckland University, Auckland, and Victoria University, Wellington) and Australia (at the Goethe-Institut, Melbourne) in combination with an open call for eco-smartphone films. *24 Frames 24 Hours* as much as *#Nucleus* embraced circularity within the production process as much as collaboration by the community that was formed for this project. This complexity links back to 'larger systems of social meaning' (De Michiel and Zimmermann 2017, 107) such as ecology for *#Nucleus* and digital literacies for *24 Frames 24 Hours*. A key focus within this research was to demonstrate that creativity ignited engagement across countries and time zones. This is further defined through composting, which 'assumes new significance in systems where meaning-making revises and subverts dominant discourses' (De Michiel and Zimmermann 2017, 107). Collaborative smartphone films provide an alternative positioning and creates a discussion based on the online and offline 'empathic' connections of the participants or peer-audience. Zimmermann and De Michiel describe this Open Space Documentary framework as an evolving transformation and transition, which is at the core of this chapter and this book as a whole. Smartphone filmmaking can change viewpoints, deadlock positions and turn conversations into a transformative force. Smartphone filmmaking can give a voice to communities and engage them in meaningful experiences. The level of collaboration can include the production of stories such as in *24 Frames 24 Hours* and *#Nucleus* or a co-creation process with communities such as *Tales from Yarriambiack Shire*. Smartphone filmmaking does not mean that filmmakers can do work for free but provides new opportunities for DIY (do-it-yourself) and zero budget productions. Cost in smartphone filmmaking is not stopping creativity; the only limitation is your imagination.

Tales from Yarriambiack Shire[8] showcased stories of rural communities. Through creating short smartphone documentaries, the community could share their stories via social media and introduce their remote communities online. These five short documentaries (total screening time of 30:15) and Cinematic VR documentary (11:12) were co-created by Schleser and Davis with the interviewees. The research implemented the co-creation process in the story development and decision-making process from pre-production, including the choice of locations to post-production and final cut. Through creating a short story format, a web series for a community could be established. On YouTube, these videos can be watched either individually or as one playlist thus providing a webisode format for rural communities (Figure 6.10).

Next to emerging collaborative storytelling practices, it is notable that the smartphone filmmaking community internationally is quite open about sharing their ideas and approaches. Several smartphone filmmakers and storytellers create online courses and training videos, such as Eliot Fitzroy for FiLMiC Pro. He is a mobile video enthusiast – voice of *Epic Tutorials*,[9] which provides a valuable resource for anyone new to the smartphone filmmaking world and those who are keen to keep up to date. Another great

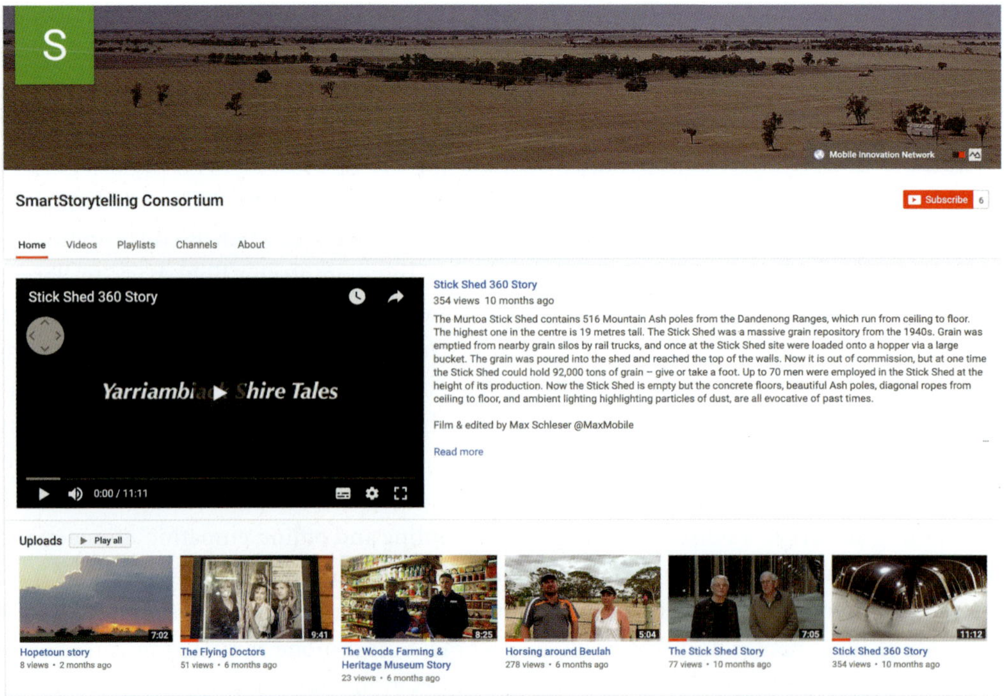

Figure 6.10 *Tales from Yarriambiack Shire* YouTube channel (Schleser and Davis 2019).

resource is Rob Montgomery's YouTube channel[10] or his *Smart Film School.*[11] Smartphone film festival director Susy Botello, who is interviewed in Chapter 3, runs the *Mobile Film School.*[12] The Mobile Motion Festival features a great blog about the latest smartphone films and developments in the world of smartphone filmmaking.[13] Filmmakers, entrepreneurs and consultants Benoît Labourdette and Jason Van Genderen lead workshops and professional development with a focus on smartphone filmmaking. Many other smartphone filmmakers also provide training courses for various industries and NGOs. Smartphone filmmakers also share their insights on their websites, such as Cassius Rayner's Budget Gear[14] overview or Richard Lackey's website, which provides an excellent resource for cinematic smartphone video.[15] The Shoulderpod website features *The Mobile Trainers World Catalogue.* FiLMiC Pro and Luma Fusion initiated the #MobileCreatorSummit.[16] Summit Producer Glen Mulcahy teamed up with speaker Courtney Jones to create the *Mobile Creator Podcast,*[17] with a focus on industry projects and productions. MINA runs workshops for community groups and professionals.[18] Max Schleser conducted workshops for various audiences and presented these at Columbia 3.0 (Bogota, Columbia), Pause Fest, St Kilda Film Festival (both Melbourne, Australia) or Doc Edge Storybox (Auckland and Wellington, New Zealand) among other festivals. In the context of community engagement, he developed a model including youth leadership training as in the *Spirit of Rangatahi*[19] or New Zealand Health Council and Ministry of Social Development (New Zealand/Aotearoa) (Davis, Waycott and Schleser 2019) (Figure 6.11).

Figure 6.11 Max Schleser's smartphone filmmaking workshop at the Festival de La Imagen in Columbia (Schleser 2015).

6.3 Smartphone filmmaking–specific formations

This section highlights how mobile screen–specific film forms not only further refine smartphone filmmaking, such as vertical video and smartphone music video production, but also have an impact on filmmaking generally. The notion of collaborative storytelling, as outlined in the last section, is further explored. The notion of accessibility facilitates experimentation and creatives outside the film industry as much as experienced filmmakers now have access to an experimental screen production space outside the traditional studio or can produce work outside the private enterprise or film commission funding structures. This allows for more stories to emerge and engage communities in novel production forms and formats.

6.3.1 Vertical video

Internationally one can observe the emergence of a vertical video streaming service, such as Quibi in the United States, which has signed Soderbergh and gained investments of Hollywood studios (Spangler 2019). The German broadcaster ZDF launched Vertical Motion Studio and in Russia the vertical feature film *V2-Escape From Hell* is in production at the time of writing this book. Next to the vertical, which defines its departure from traditional cinematic aspect ratio into new domains, one can observe the formation of a new profession of screen producers, the online and digital video creators at congresses such as VidCon (including the 'Stars of Tik Tok'), new production hubs such as YouTube Spaces and new apps/software such as Adobe's new editing software Adobe Rush. Rush was created for online videographers and smartphone filmmakers to edit on the go and can be vertically used for editing.

The *Vertical Video Syndrome* generated a substantial YouTube viral success, and, in the past, some apps would change vertical video to cinematic 16:9 aspect ratio, but within a short time this perception seems to have changed significantly. This development was driven mainly by the smartphone-users or rather makers and caught the interest of independents and industry alike. In 2013 Vertical Cinema launched in Austria, in 2014 the VFF (Vertical Film Fest) in Australia and in 2019 SLIM in New York, United States, and Verti Films in Prague, Czech Republic (Figure 6.12). Prior to that Miriam Ross created the *Vertical Cinema Manifesto* in New Zealand in 2013. Moreover, she has contributed to creating an understanding about this format through her work on Rhizome (Ross and Glen 2014) and in *Mobile Story Making in an Age of Smartphones* (Ross and Neal in Schleser and Berry 2018). Ross points at the blurring of the boundaries between 'amateur' and professional media. She outlines what was previously positioned in the realm of vernacular practices is now featured in corporate environments. 'However, recognising the slippage between amateur and professional distinctions does not mean underestimating the agency users have to work these contexts in innovative and challenging ways' (Ross 2018, 158). A vital example of this merging of distinctions is the vertical smartphone film *POCKET* by Mishka Kornai and co-directed by Zach Wechter. The growing up and first engagement with sexuality by American teenager Jake Tillner (played by Mace Coronel) is revealed

Figure 6.12 Verti Films Prague 2019. Festival screening Convent of St Agnes of Bohemia in Prague. Vertical film Kokoro by Ondrej Bojo. Winner Tall Idea Prague 2019 Award (Verti Films 2018).

through the interface and point of view of the iPhone. Using a similar creative strategy like the *Life and Death of the iPhone*, the 17-minute vertical video provides an authentic account and perspective into the life of a teenage boy. The narrative peaks in the climax of an IRL (in-real-life) encounter with his Instagram crush Farrah, revealing his insecurity without a curated and crafted social mediated world. The film can be seen as a deep dive into the world of Generation Z and their public intimacies and social insecurities.

The argument is not that vertical will establish itself the new standard for the entire mediascape, but we can observe a proliferation in the online domain. Especially new mobile social-media video sharing platforms such as Periscope, Snapchat, Facebook Stories, Instagram Stories and TikTok drive this development. The latter is gaining substantial attraction, especially with younger audiences. Transmedia storytelling approaches have successfully demonstrated how to integrate these viral videos with more traditional cinematic and broadcasting formats. Another key characteristic of most vertical videos is their microformat, Snapchat and TikTok videos are maximum 60 seconds, Facebook Stories 20 seconds, Instagram Stories 15 seconds and Quibi streaming service will provide content in a vertical and horizontal format in under 10-minute formats.

6.3.2 Music videos

With arts funding for mobile moving-image arts and experimental film becoming more and more difficult to secure, music videos provide a more accessible avenue for experimentation. Several musicians have sized the day and utilized the affordances of smartphones in terms of their mobility, accessibility and ubiquity for their music videos. With more bands being able to produce music videos, this is an opportunity for bands as much as filmmakers to 'stand out' through these novel and creative approaches rather than relying on big budgets. In combination with online dissemination, video platforms and social media, smartphone filmmaking's connectivity provides opportunities to engage with fan bases.

When Coldplay performed at the Global Citizen Festival in Mumbai in 2016, they asked fans to upload their videos to Instagram. Their music video *Amazing Day*[20] was a compilation of vertical or rather 1:1 video. Other musicians who decided to choose vertical video as their preferred choice of format include Saweetie's *B.A.N. (Official Vertical Video)* or Halsey's *Without Me (Vertical Video)*. In the context of music videos being produced on mobile devices, one could point at the US music band Presidents of the U.S.A. They created their 2008 music video *Some Postman* on multiple mobile phones in the production company Film Headquarters studios. The music video reveals the montage approach that also artists like Steve Hawley's in *Speech Marks* (2004) applied. *Speech Marks* was exhibited in the earlier mentioned FIL*M*OBILE exhibition. The moving-image artwork uses a montage concept and stitches the 3 gp video files in their original resolution (176 × 144) into a 1280 × 720 (720 p) video format. The mobile art video references the every day and produces a montage from multiple perspectives in a diary format, filmed over one week in 2004 (Figure 6.13).

The Indie artist Grimes and HANA filmed seven music videos during their European Tour in 2016 on smartphones. Using some mascaraed, costume and most significantly iconic locations the music video could be produced on the go in between their concerts.

Figure 6.13 Still image: *Speech Marks* (Hawley 2004).

Grimes's brother Mac Boucher points to the key creative strategy of improvisation, which he describes as 'for all its temporary limitations, the positives far outweigh in terms of creative freedom for we were only ever considered insane tourists with a weird selfie stick and occasional blasphemous short shorts and robes' (Leu 2016). The guerrilla-style filmmaking gave Grimes the creative freedom she missed in studio shoots. In an interview in the fashion magazine *Elle*, she points out that working without the constraints of the production schedules in a studio, 'You don't really get to improvise or have any creative freedom by the time you are actually shooting' (Tang 2016). These surreal-themed music videos are driven by the artist's emotions and their encounters with the location.

Another filmmaker whose music videos are defined by an experimental approach is Luis Eduardo Juarez. His music video *Fly Away* for the L. A. reggae band Christafari screened at MINA in 2014. His experimentation with manipulating time, through playing sequences backwards in the music video, creates fluid and bizarre time-space continuum (Figure 6.14). After graduating from the New York Film Academy in 2008, he found Indiefone, one of the first iPhone Film Festivals. Currently, Luis Eduardo Juarez is working on #ScratchTheFate, a feature smartphone film. In this film, a Latino protagonist is faced with a challenging situation when he wins a lottery ticket but does not have social security number and thus not being able to secure his lottery but is also declared an illegal alien (Figure 6.15).

Figure 6.14 Still image: *Fly Away*. Luis Eduardo Juarez, Mark Mohr and Avion Blackman (Juarez 2014).

Figure 6.15 Still image: *#ScratchTheFate*. Luis Eduardo Juarez and actor Daniel Cadena (Juarez 2020).

In 2014 Juarez explored underwater smartphone filmmaking and created *Oceans (Where Feet May Fail)* for Christafari. The music video features several underwater shots using water as a VFX component. Using slow motion and filming in a higher frame rate, the underwater effects could be enhanced (Figure 6.16). Beyoncé, Eminem, John Legend and Edward Sharpe and the Magnetic Zeros are further examples of music videos produced using smartphones. While these examples illustrate the engagement of major labels and entertainment companies in smartphone filmmaking, it also emphasizes the recognition of smartphone filmmaking beyond the independent sector. In Beyoncé *7/11* she filmed herself and her crew in a selfie-style, which underlines the intimate and personal qualities of smartphone filmmaking in comparison to a full-fledged studio or on-set music video productions. Google sponsored Eminem's *Venom* production, which was filmed on the Pixel 3. The production featured a film crew using the mobility of the smartphone to its advantage when filming in the Empire State building.[21] Another Google Pixel (2) production is the music video *A Good Night* by John Legend, which was directed by Mishka Kornai who co-directed the earlier mentioned smartphone film *Pocket*. The music video production featured twelve phones to create the *Matrix* like bullet-time effect.[22]

Similarly, the cast and crew list in *No Love Like Yours* for Edward Sharpe and the Magnetic Zero's music video is almost comparable to a feature film production including performances with eighteen dancers and a substantial crew for the art department. By means of shifting expenses from camera rentals, filmmakers realize creative projects and have a license for experimentation. It is also great to hear the filmmakers commenting in their behind-the-scenes videos on the democratizing aspect of smartphone filmmaking[23] (Figure 6.16).

Figure 6.16 Production stills: *Oceans*. Luis Eduardo Juarez and Makamae Kailani (Juarez 2014).

6.4 Smartphone filmmaking modes

In order to illustrate that the smartphone filmmaking modes outlined in Chapter 1 find a wider resonance than the MINA screening and festival, this section reviews these modes in the contemporary mediascape. The difference to the filmmakers, artists and creative technologies featured in this book is that they developed a substantial and critical practice around creative mobile media. In contrast, most of the following filmmakers made an excursion into the world of smartphone filmmaking and then returned to more 'traditional' industry practices and cameras.

6.4.1 'Red carpet in your pocket' mode

In 2011 the feature-length mobile film *Olive* was released by Hooman Khalili, which he chose to film on a Nokia N8 smartphone.

While Khalili used more of a 'traditional' production format, Fosheim in *Uneasy Lies the Mind* experimented with the iPhone's original look and feel in relation to traumatic experiences and fragments of memory. In *I Play with the Phrase Each Other*, Jay Alvarez constructed the entire narrative around mobile phone calls between actors and Alvarez himself, in the feature-length black-and-white film. Michael A. Cherry filmed *9 Rides* in 2016 which follows an Uber driver on New Year's Eve using an iPhone 6s in 4K. Another example of the director and cinematographer combination is Michael A. Cherry and Ricky Fosheim's *Uneasy Lies the Mind* (2014), which is a feature-length psychological thriller about a couple spending an evening in a winter mansion when an old friend visits them and the evening takes an unexpected turn (Figure 6.17). In a similar approach as in *Unsane*, the iPhone 5 look supports an unsettling story. Ricky Fosheim used the app FiLMiC Pro

Figure 6.17 Still image: *Uneasy Lies the Mind*. Actor Isaac Nippert (Fosheim 2014).

and used a f1.4 Nikon Nikkor AI-S F-mount lenses and shot the majority of the movie with 35- and 50-millimetre lenses.

> From the start, I welcomed grain, dirt, flares and any other techniques that would help me create a dirty, fragmented and organic look . . . . The iPhone footage was raw, dirty, vignetted and unlike anything I'd seen before. I immediately fell in love with the look, and I decided to fully embrace these unconventional limitations as powerful storytelling tools.
>
> (Fosheim 2014)

Uneasy Lies the Mind experiments with 6 fps (frames per second) to capture fragmented memories, while otherwise shot in 50 Mbps instead of the iPhone's native 24 Mbps (Figure 6.18). The H.264 footage was transferred into ProRes 4:2:2 and Colorist Brent Greer applied Blackmagic Design's DaVinci Resolve for balancing and matching shots as the iPhone's textured and vignette image was maintained to visually support the story world of envy, secrets and paranoia.

The film was produced in one location, which adds to the smart storytelling approach that is also realized in *Blue Moon*. Stef Harris is a full-time police officer, who had written novels and adapted them into successful feature films before. He produced this thriller in one week with antagonist Mark Hadlow and protagonist Jed Brophy on location in the small town Motueka on the South Island of New Zealand. Cinematographer, Ryan O'Rourke, also a police officer, used FiLMiC Pro and Moondog panoramic lens. Supporting roles were also casted from the police force who played as police officers in their natural habitat but had no formal acting experience or training. By means of using a smartphone,

Figure 6.18 Production still: *Uneasy Lies the Mind*. Jonas Fisch and Michelle Nunes get ready for a take as Ricky Fosheim operates camera (iPhone 5) (Fosheim 2014).

Stef Harris could hide the camera in order not to intimidate them. The crew had access to the petrol station for five hours from midnight for six nights to film the drama production. Stef Harris's presentation, 'The creation of an 86-minute noir thriller movie on iPhone 7+ in one week with a budget better suited to purchase a high mileage Honda Civic', at the International Mobile Storytelling Congress outlined his creative strategies to realize this passion project. He gained support from the small town via a Facebook page and could use all his experiences as a police officer to fine-tune the script which took him about a year after signing the actors, who worked on the project without receiving payments. Other crew included Ben Dunker, a sound recordist with Hollywood credits, Dan Hennah, art director and production designer with credits on the *Lord of the Rings*, and make-up artist Shawn Foote from Weta Workshop among others who volunteered their time. The iPhone's flexibility and mobility facilitated the production process that was required by Stef Harris and his crew. *Blue Moon* screened at more than twenty festivals and won several awards including distribution deals and funding by the N. Z. Film Commission (Figure 6.19).

Other smartphone films in this category are *Snow, Steam, Iron* by Zack Snyder (2017). The Hollywood director who is known for features like *Batman v Superman: Dawn of Justice* was also the director of photography and thus another example of the director-cinematographer. In the same year, the comedy feature drama *High Fantasy* directed by Jenna Bass was filmed in South Africa. During a camping trip, the protagonists' bodies are switched, which allows the filmmaker to navigate gender and race roles and stereotypes: 'the stage is set for comedy to turn to tragedy, for the fantasy of South Africa's "Rainbow Nation" to become a painful awakening' (High Fantasy 2017).[24] *Char Man*, a cross between a road movie and a horror drama, smartphones and mobile devices (in the form of iPads), was the ideal choice of camera for the DP and assistant camera operators in this feature film production. The film was shot in two and a half days with the directors and producers, Kurt Ela and Kipp Tribble, doubling as assistant camera operators and DP Nick Greco

Figure 6.19 Production still: *Blue Moon* (Harris 2018).

using an iPad, who has several acting, producing and writing credits, could also be turned into cast when needed.[25] These two mobile and smartphone film productions demonstrate that for a road movie or mockumentary concept, the mobility and spontaneity of a smartphone makes it the ideal camera for independent productions.

6.4.2 Conversational mode

As earlier described, the conversational mode can be seen as a synthesis of smartphone filmmaking with mobile social media. As much as likes and comments engage in online conversations, films in the conversational mode make use of the network media platforms and new dissemination formats to engage audiences in novel experiences. These episodic forms can be linear but work in micro-units that connect to a larger story world or project in a transmedia approach. This mode also provides means to open up conversations with the audience.

In 2016 Hannah Macpherson produced the *Sickhouse* experience using Snapchat. In a *Blair Witch Project* style film, a group of teenagers live stream their journey into the woods to explore the haunted house, Sickhouse. The production by Indigenous Media was live broadcasted via Snapchat and then reedited.[26] In the context of factual filmmaking *#FollowMe*, the Dutch broadcaster VPRO illustrates how online conventions and aesthetics can be integrated into a broadcast documentary. *#FollowMe* reveals how bots, comment-on-demand and fake accounts drive the attention economy on the social media platform Instagram and exposes its unregulated dark side. While it is not filmed entirely on smartphones, the documentary demonstrates new opportunities for audience engagement and dissemination via Instagram[27] and YouTube.[28] *#FollowMe* was released via Instagram, YouTube and broadcast in the Netherlands. Here one should mention that the latter version is displayed in the vertical format. Filmmaker Nicolaas Veul (@nicolaas_veul) reveals how likes and followers can be manipulated in this new social media industry of influencers (Figure 6.20).

Another prominent example of the conversation mode combining smartphone filmmaking, social media and storytelling is *Eva.Stories*. Produced in 2019 by Mati Kochavi and his daughter, Maya Kochavi, they chose Instagram as the main dissemination platform in their attempt to engage young audiences into the subject of the Holocaust. *Eva.Stories* recreates the diary of a teenager through the perspective of the contemporary Instagram aesthetics that is staged in 1944. The Instagram story was produced with the budget of a feature film and used an art department to recreate the time of fascism and invasion of the Nazis in Hungary. The juxtaposition of hashtags, emojis and Instagram graphics brings the genocide into a contemporary social media world and reveals that history from the perspective of thirteen-year-old Eva Heyman. Another Instagram story that engages in history is the interactive time travel project *Throwback '89* by German public-service broadcaster, ARD's news *Tagesschau*, which is dedicated to the Fall of the Wall on 3 October 1990. The episodic online experience successfully merges historical TV archive material and fictional personal diary film account. The Instagram episodes are between one and three minutes and were released via the news TV show's Instagram channel. The Instagram clips merge news from the reunification of the DDR (German Democratic Republic) and BRD (Federal Republic of Germany) through the *Tagesschau* archive and a fictionalized Instagram story of teenager Nora in the DDR in 1989. In the context of smartphone

Figure 6.20 Still image: *#FollowMe* (VPRO 2019).

filmmaking and broadcasting, one can also point at the work of SWR-Reporter Jürgen Rose. In 2013 he wrote and directed *Hungerlohn am Fließband: Wie Tarife ausgehebelt werden* (= *Starvation Wages: How Pay Scales Are Cancelled*) filming under-cover in a Mercedes-Benz factory. Jürgen Rose took temporary work with an agency ending up working in a factory alongside contract workers to worse conditions. He documented his experience in a diary format. Next to generating discussions online, smartphone filmmaking can operate in environments that may otherwise be either difficult to operate cameras or unable to gain permissions for. Another key example in this context is *This Is Not a Film*[29] (2011) by Jafar Panahi. The Iranian director was on house arrest and produced a video diary during this time. The project was released in 2011 in Australia and also smuggled to the Cannes Film Festival. Panahi was awarded the Carrosse d'Or in his absence.

In 2009 Sepideh Farsi filmed *Tehran without Permission* (*Tehran Bedoune Mojavez*). As the name indicates, she had no release forms and production documentation (Handke 2009). This documentary film can thus reveal the social and political tensions of a city in transition in-between tradition and modern, rich and poor, public and private life in Tehran. These examples demonstrate that when no camera teams are present, smartphone filmmakers can capture events and stories, giving a voice to communities.

6.4.3 Participatory and engagement mode

The documentary *#MyEscape* by German broadcaster Deutsche Welle captures the journeys of refugees on their way from Syria, Afghanistan and Eritrea to Germany on their smartphones. The first-person accounts reveal the refugees' departures, their old and new homes, their personal situations, their memories and traumatic experiences.

In 2014 Ossama Mohamed and Wiam Simav Bedirxan produced *Silvered Water: Syria Self-Portrait*. Mohammed edited the material, which Simav shot herself in the besieged city with excerpts from '1001' mobile phone and smartphone videos of heavy shelling and aerial bombardments.[30] Peter Snowdon's *Uprising* is a feature-length documentary composed entirely of videos made by citizens and residents of Tunisia, Egypt, Bahrain, Libya, Syria and Yemen. The film uses this footage not to recount the actual chronology of events or analyse their causes but to create an 'imaginary pan-Arab uprising that exists (for the moment) only on the screen' (Snowdon 2014).

6.4.4 Smartphone native mode

This mode shares similarities with MoJo, mobile journalism, such as the work of *#HashtagOurStories*[31] or prominent TikTokers such as the *Stars of Tik Tok* at VidCon.[32] Next to the earlier mentioned 'Made on Mobile' award at the SmartFone Flick Fest, FiLMiC Pro's Filmic Fest featured a made-on-mobile category.[33] As part of the interview for this book, Felipe Cardona was asked to share his iPhone production studio. These screenshots provide an excellent overview of the possibilities for working exclusively on mobile devices and smartphones. With increased computing power in smartphone and mobile devices, a growing number of apps for post-production such as greenscreen production and VFX (special effects) are available on the App Store or Google Play (Figures 6.21–6.24).

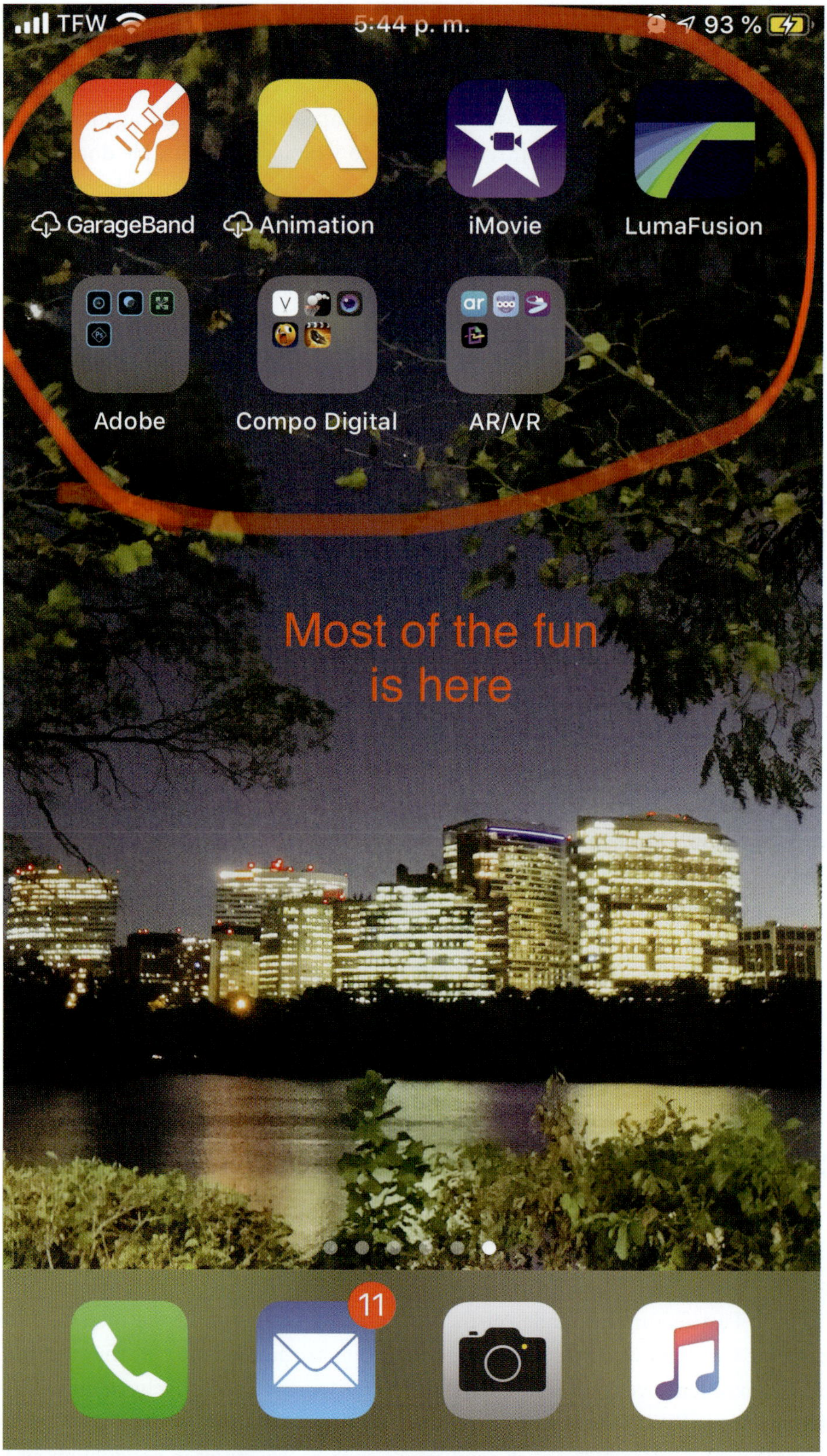

Figure 6.21 Smartphone filmmaker Felipe Cardona's apps (Cardona 2019).

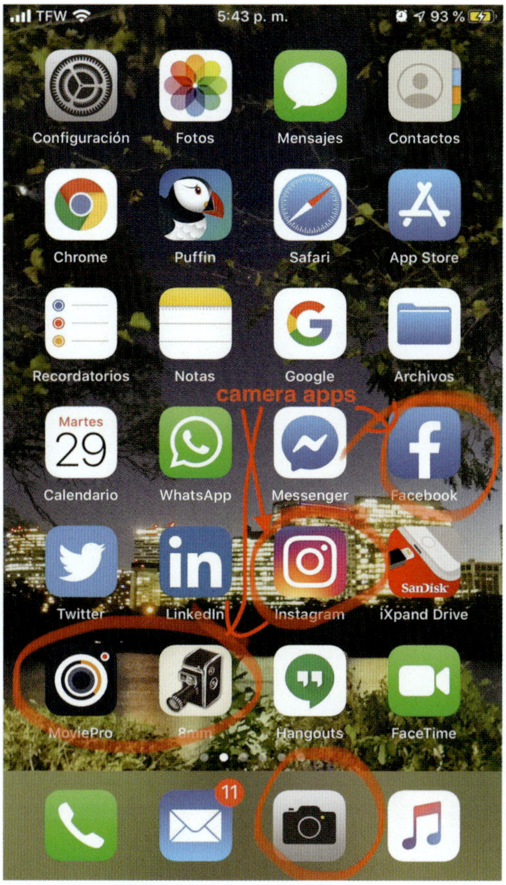

Figure 6.22 Smartphone filmmaker Felipe Cardona's camera apps (Cardona 2019).

The experience of editing on a mobile device can be compared to painters working with their canvas on location in a landscape setting rather than in the studio. Angelo Chiacchio sees it in a similar way, 'it is like painting a landscape or a portrait while it is alive and moving instead of doing it from a photograph' (2020). He created *Ephemera*, and the title reveals both the subjects of his project and his workflow. In 2017 he visited evanescent places and cultures while editing and publishing from mobile devices during a 300-day journey (Figure 6.25).

> An image on Instagram or video on Facebook lives truly for a few moments before getting lost in an infinite scrolling feed of attention. However, a little bit like a postcard that usually ends up lost in a dusty drawer, if the message conveyed is important, it will probably be cherished by the receiver and kept in a special place.
>
> (Chiacchio 2019)

Next to video production on smartphones, music production has seen a similar proliferation. In November 2018, attendees of MINA Smart Storytelling Symposium were given access to a selection of some of the earlier submissions to a compilation album showcasing contemporary trends in mobile music-making (Figure 6.26).

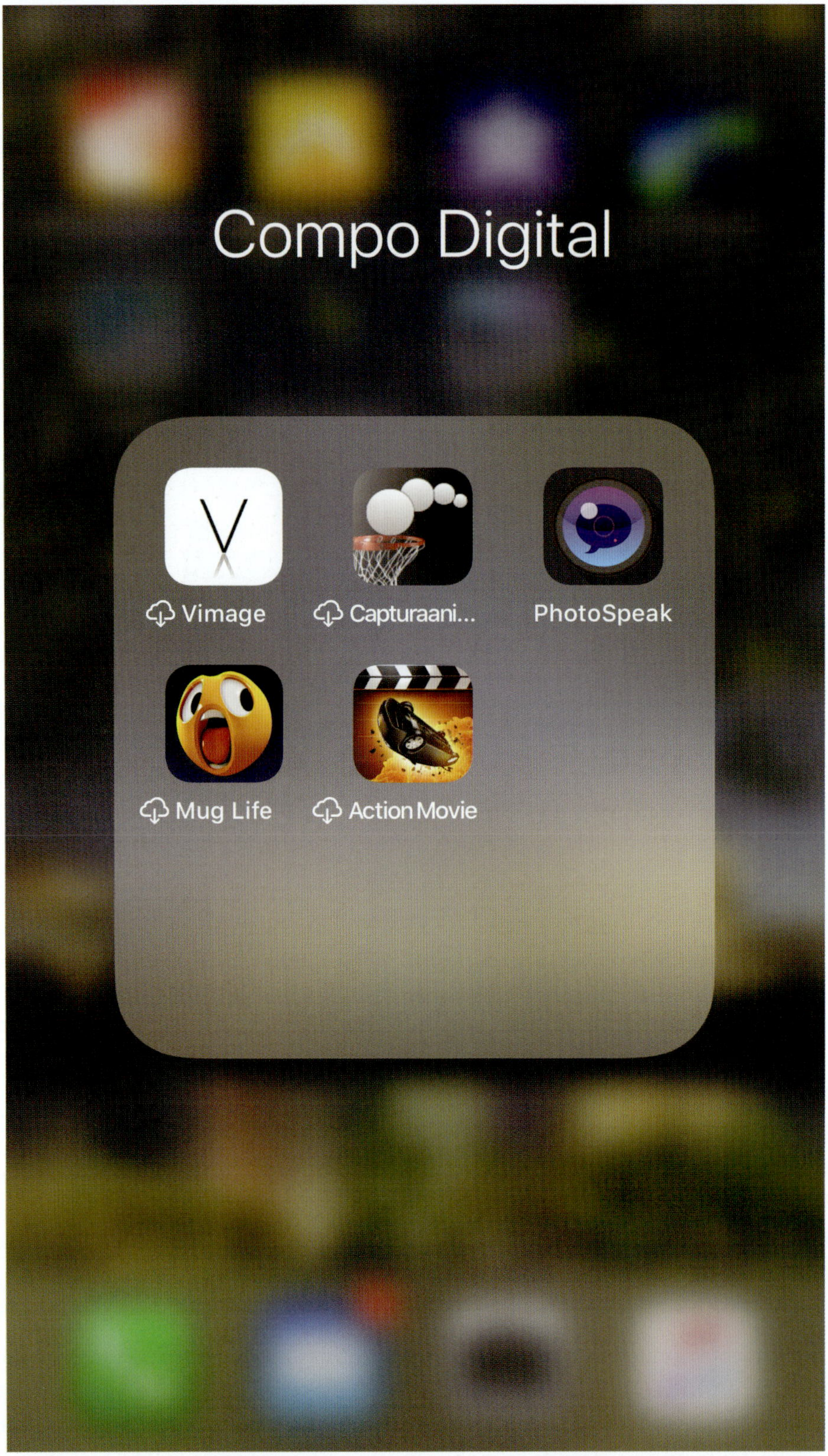

Figure 6.23 Smartphone filmmaker Felipe Cardona's VFX apps (Cardona 2019).

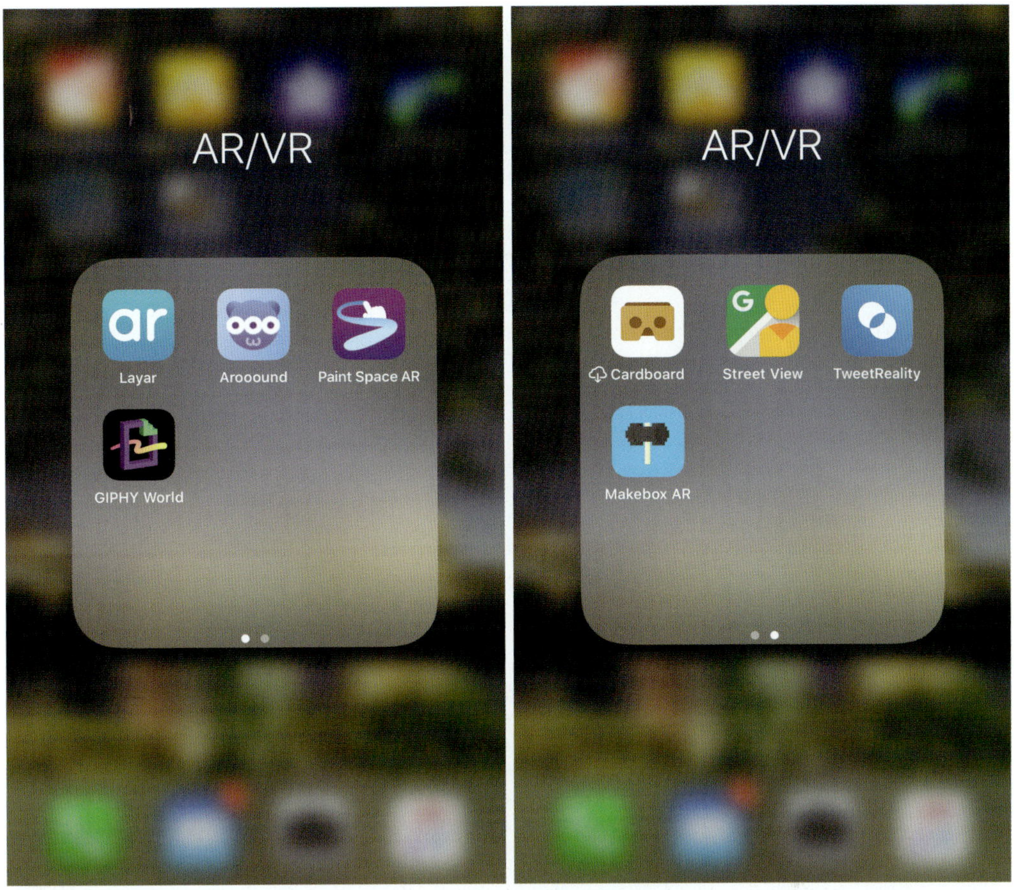

Figure 6.24 Smartphone filmmaker Felipe Cardona's AR/VR apps (Cardona 2019).

The album was produced and curated by the MINA speaker Martin Koszolko and released in October 2019 under the title *Mobile Strategies: Battery-Powered Sonics* on the Australian label Clan Analogue. The album includes twenty-five international artists who use a broad range of mobile technologies to produce various forms of electronic music.[34]

6.4.5 Poetic and experimental mode

In 2011 Korean filmmakers and brothers Park Chan-wook and Park Chan-kyong premiered the fantasy-horror short film *Night Fishing* at the Berlinale. The imperfect smartphone aesthetic suited this fairy tale about death and reincarnation. *Night Fishing* (*Paranmanjang*) dives into a story world of a night fishing session and a metaphor for reincarnation. A fisherman catches a fish, which transforms into a young woman and then merges into a ritual of submersion. She wears a funeral dress and calls the protagonist 'daddy'. The brothers Park Chan-wook and Park Chan-kyong merge western and eastern genres and filmic conventions (Figure 6.27). The iPhone 4 aesthetic adds to the surreal environment filming with soft-focus, extreme camera angles and featuring a music performance in the

Figure 6.25 Production on the move *Ephemera*. Angelo Chiacchio editing on location (Chiacchio 2018).

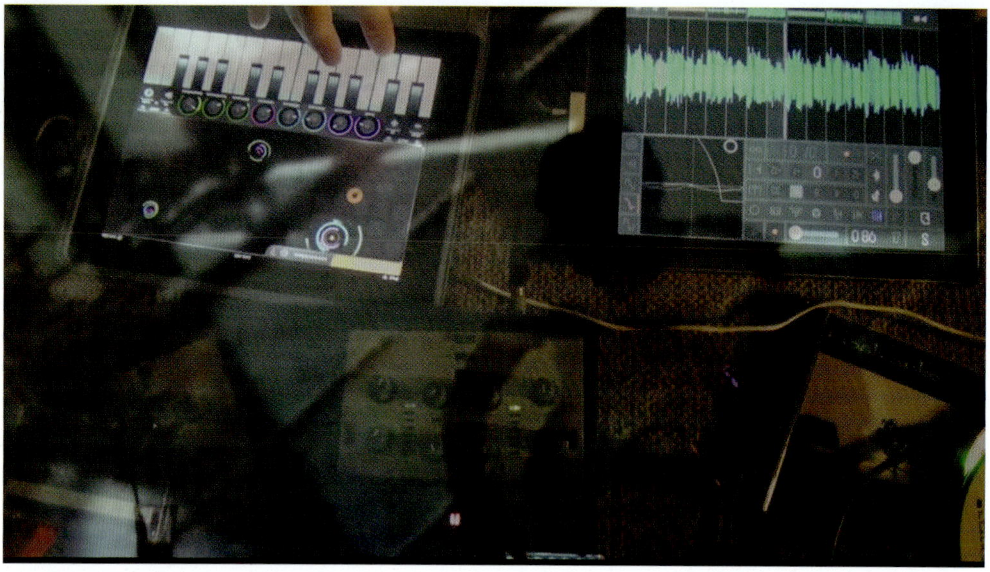

Figure 6.26 Mobile music production on the move by KOshowKO (Koszolko 2019).

beginning. During the night fishing sequence, the film shifts from colour to black and white and then fades back into colour symbolizing past and present. Most sequences are filmed handheld. The unstable nature of this technique supports the film's unsettled look and feel. According to *Screen Daily*, the Park brothers could secure a substantial budget for this short film ($130,000) and well-known cast and musicians (actor/pop star Lee Jung-hyun and a soundtrack by UhUhBoo Project) (Paquet 2011). More significantly the smartphone film not only received mention in industry trade press but was awarded in

Figure 6.27 Still image: *Night Fishing* by Park Chan-wook and Park Chan-kyong as PARKing CHANce film (Park Chan-wook and Park Chan-kyong 2011).

Figure 6.28 Production still: *Night Fishing*. Filming of the opening sequence using iPhone 4 (Park Chan-wook and Park Chan-kyong 2011).

2011 with the Golden Bear for the Best Short Film at the 61st Berlinale International Film Festival (Berlin, Germany) (Figure 6.28).

Luca Lancise's *Moravia Off* combines archive material, and personal video-diary style extracts about the Italian writer Alberto Moravia and his journey around the world. The experimental smartphone film screened at the Rome Film Fest (Festa del Cinema di Roma). The audio-video installation *Ah humanity!* by Verena Paravel and Lucien Castaing-Taylor is filmed following the Fukushima nuclear disaster in Japan. In order to film the devastated landscape, the filmmakers used a telescope on a camera of a mobile phone resulting in a 'randomness and limiting the possibilities of control over the image'.[35] The work is distributed by LUX[36] and is presented as a large-scale audio-video installation (Figure 6.29).

Hope Tucker's *Puhelinkoppi (1882-2007)* is a documentation of the last public phone booths in Finland. Using a split-screen, the filmmaker combines personal observations with facts about the telephone system in Finland. The split-screen merges into a single screen work showing the removal of the last phone booths, which is filmed on a Nokia phone. A decade after the production, this becomes even more interesting as the Finnish mobile phone company Nokia only plays a minor role in the smartphone market and could become as redundant as public phone booths (Figure 6.30).

Figure 6.29 *Ah humanity!* Installation (Paravel and Castaing-Taylor 2015).

Figure 6.30 Still image: *Puhelinkoppi* (1882–2007) (The Obituary Project 2010).

British filmmaker Scott Barley's *Sleep Has Her House*, *Hinterlands*, *The Green Ray* and *Womb* are imaginative journeys similar to hypnotic experiences. He mixes live-action, photography and hand-drawn images. Using abstraction and working within structuralist and surreal traditions, he illustrates the low-light capacity of smartphones imaginatively and successfully. His cinematic landscapes use nature as a layer that is superimposed and composited into subtle experiences in dark environments. The choice of colour is synched with the reverberation creating a space between cinema and the uncanny. Since early 2015, Barley has exclusively shot his films on iPhone (Figure 6.31).

Another British filmmaker working with smartphones is William Brown. His essay films include *Letters to Ariadne*, *#randomaccessmemory*, *Vladimir and William* and *The New Hope*. He is combing the essay film and documentary approaches intertwined with fictional elements reflecting on love, cinema and the 'image virus'. The latter is the key theme of his smartphone film *Selfie*, which deconstructs contemporary addictive image culture. In his voice-overs, his cinematic criticism and provocations vocalize how smartphone filmmaking can create an alternative discourse and intervention to the established film industry ('just because the image resolution is good does not mean that the film is good', says Brown [2014]) and looks for the deeper meaning behind the moving-image world and its representations (Figure 6.32).

Figure 6.31 Still image: *Hinterlands* (Barley 2016).

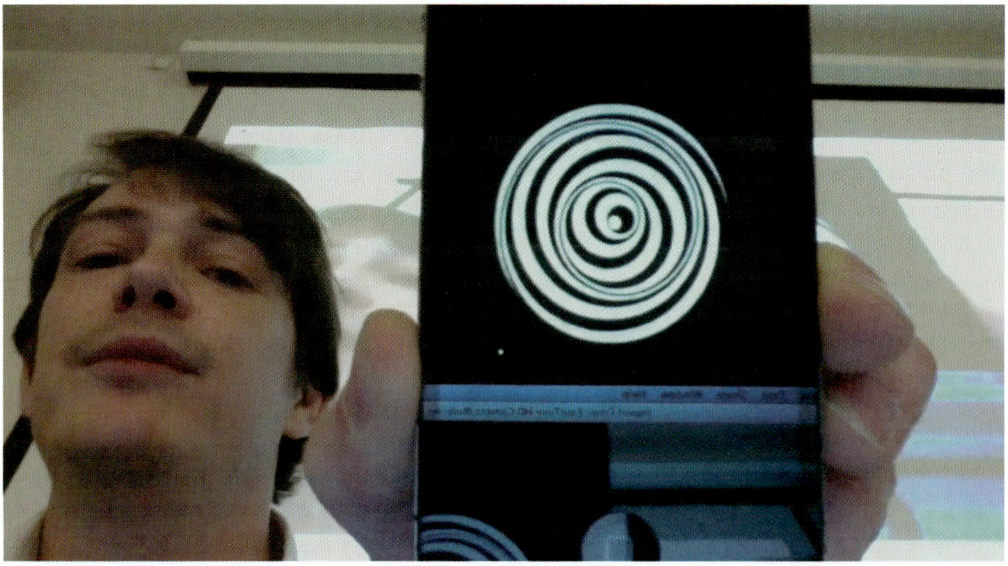

Figure 6.32 Still image: *Selfie*. William Brown showing Anémic Cinéma (1926) on a smartphone as a metaphor for the mesmerizing 'image virus' (Brown 2014).

6.5 Mobile XR

This section will explore some of the most recent developments in smartphone filmmaking and speculate on its creative futures. Smartphone filmmaking is not only a fertile ground for aesthetic experimentation but also acts as an innovation sandbox for screen production and adjacent disciplines. Mobile media and smartphones filmmaking as a scholarly field of study is situated within film, TV, digital media and screen production. In the near future this expanding field will increasingly intersect with ubiquitous media, the internet of things, apps, social and network as well as wearable media (Figure 6.33). Smartphones are a median that interconnect with other devices and data. As an example, Apple Watch with the FiLMiC Pro app or the Sony Cyber-Shot QX10 lens comes to mind. The latter connects to smartphones via WiFi and allows the mobile lens to be placed or held in any position possible and can be remotely controlled via the smartphone interface. The mobile phone, cell phone as much as the smartphone are a fusion of lens-based media and communication media for the very first time as a ready off the shelf product. While the first camera phone, the Sharp SH-04, was introduced as a marketing idea in Japan in 2000 (Schleser 2014, 156) and lead to a proliferation of experimentation as discussed in this book. Now one can observe a similar creative engagement with ubiquitous media. In 2019 two Austrian filmmakers, Torsten Frank and Marco Neumeyer working for the agency Distillery, used an Audi to film a new form of a skateboard video for skater Daniel Ledermann.

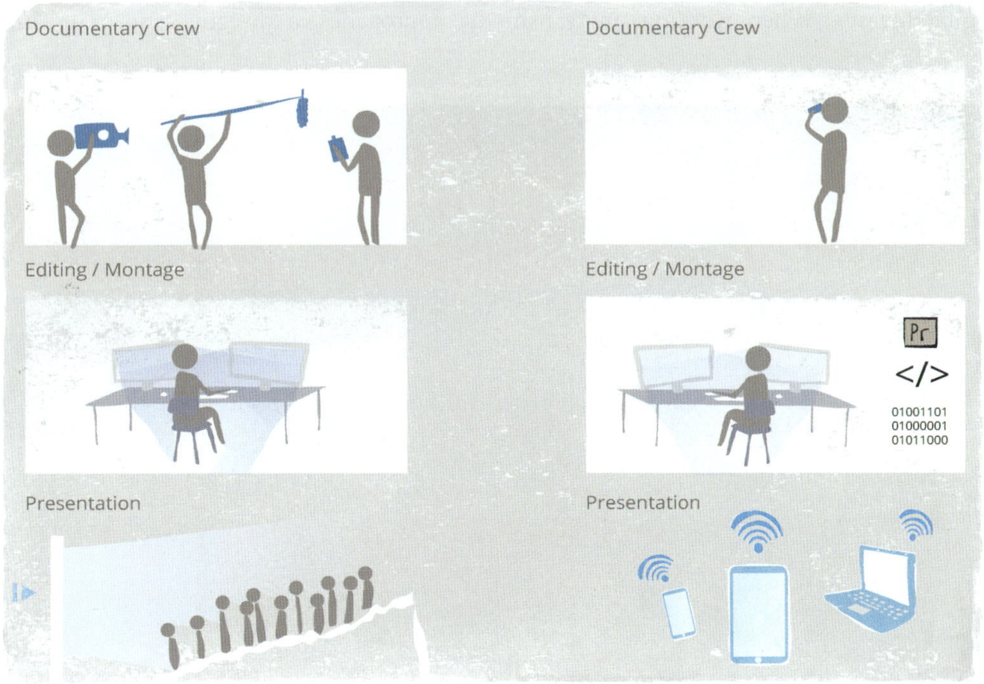

Figure 6.33 Twenty-first-century smartphone film crew (D'Oro 2020).

These cameras provide new perspectives to capture our world in novel camera angles beyond standard handheld or tripod set-ups. One of the extreme examples is Luke Geissbühler's *Homemade Spacecraft*, which screened at the 2012 International Mobile Innovation Screening demonstrating a DIY and maker approach as part of the Brooklyn Space Project. As the smartphone film name indicated, an iPhone and a GoPro camera were sent into the stratosphere with a weather balloon (Figure 6.34). Next to these creative interventions, drone cinematography evolved like smartphone filmmaking and has grown into a scene with its creative as well as commercial applications, including its own festivals, such as Peugeot Drone Film Festival[37] or New York City Drone Film Festival.[38] Similarly, robot film festivals[39] provide us with an expansive outlook into possible futures in film production.

Alongside the development of smartphone gear, such as Gimbals, drones, led lighting and wireless mics, roles of a crew and smartphone filmmakers' skill-sets developed. Several smartphone filmmakers in this book embody the one wo/men-band or work collaboratively in non-hierarchical manners. Next to the traditional roles of director, cinematographer and audio-recordist on location, which became fused into the one wo/man crew, the role of the editor in post-production can be complemented by a developer and/or creative technologist. In the domain of the producer dealing with the dissemination strategy, a social media community manager or social media content manager is required to keep the engagements and interactions active. Section 6.2 illustrates how network and social media provide an infrastructure for collaborative practices. Creativity as an engagement form provides opportunities for communities to create representations of themselves according to their own agendas. Smartphone filmmaking is becoming a key element in twenty-first-century digital literacy. When working on transmedia approaches, a designer and/or creative

Figure 6.34 Still image: *Homemade Spacecraft*. Smartphone filmmaking at 30 kilometre/10000 feet in the stratosphere (Geissbühler 2012).

technologist should be introduced into the project before pre-production and conceptualization of the narrative strategy. Audiences and participants can engage in various stages of the production process and access the story components via numerous touchpoints. Section 6.4.2., which explored the conversational smartphone filmmaking mode, illustrated these approaches to transmedia storytelling and cross-media production (Figure 6.35).

Creative Commons established itself as an alternative to integrate with existing copyright laws and create a new dimension for the online world. The smartphone ecology is positioned at the intersection of industry, alternative broadcasting, cinematic models, workflows and distribution systems while embracing new dynamic spaces and models. Currently, blockchain technology and smart contracts can provide some disruption in the film dissemination process. Here one could point at the Russian smartphone comedy film *Fagotto*. Boris Gouts shot the feature in just four days with a micro-budget in 2018 on an iPhone. In 2019 he completed another smartphone feature production *We Look Good in Death*. *Fagotto* is now distributed via Cinezen,[40] which is a Swedish start-up founded in 2017. Cinezen can be described as an independent video store that uses distributed ledger technologies (DLT). As a decentralized peer-to-peer (P2P) marketplace Cinezen offers more transparency and verifiable reporting. There is enormous potential in the direct and instant payments from end-users to content rights holders via Ethereum. CINEZEN's idea of independent video stores creates opportunities for niche markets and films to be accessible internationally. While peer-to-peer websites were highly successful, they did not operate within the current copyright framework. With smartphone filmmaking, film

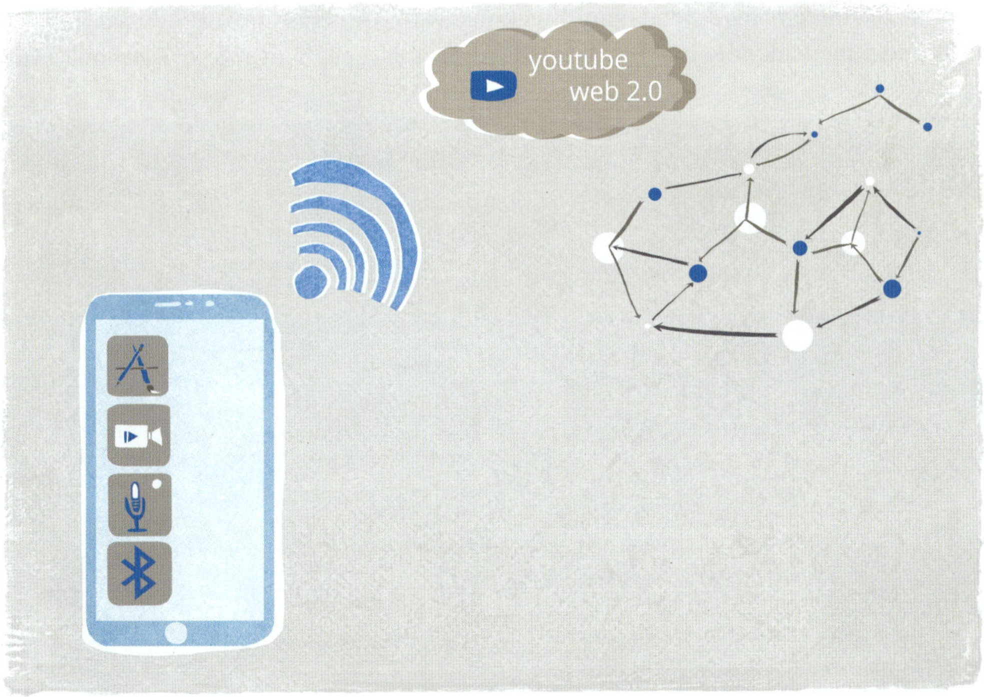

Figure 6.35 With the introduction of video platforms as exemplified through YouTube in 2005, smartphone filmmaking is increasingly disseminated via networks online, in the cloud (D'Oro 2020).

production can be seen as a more egalitarian and approachable industry. Now CINEZEN believes that 'anybody who loves and knows cinema will be able to start their own unique CINEZEN store and become a "film distributor" to their friends and followers. All store owners become social media influencers and bring their own audience to the platform (Klebanov 2020).'[41] The engagement of viewers and audiences as much as creatives and makers in the distribution platform further supports the argument that smartphone filmmaking can create a more open ecology or in the terminology of De Michiel and Zimmerman 'open space' (2017) (Figure 6.36).

Rather than focusing only on return in revenue, smartphone filmmakers work in the attention economy and provide opportunities to create sustainable connections to audiences. With an increasing outlet of screens and immersive media, storytelling and content production can take a number of forms beyond the cinema screen. The flow of stories between devices, screens and platforms is at the heart of transmedia storytelling. Urban screens as much as XR provide novel experiences and reach audiences outside the cinema and beyond desktops or living room screens. The terminology XR refers to extended reality. At the 72nd Cannes Marché du Films in 2019 Cannes XR[42] was introduced to provide a forum for immersive media including AR (Augmented Reality) and VR (Virtual Reality) into one domain. These new formations are increasingly mobile and embrace locative media.

With the recent proliferation of 360° video cameras and VR headsets as well as AR technology, which can easily be integrated into the creative workflow for storytellers and filmmakers, this chapter reviews some of the current developments in the emerging field of Mobile XR (Figure 6.37).

In the last three years, several omnidirectional video cameras were introduced, and 360° video and Cinematic VR functionality added to YouTube, Vimeo and Facebook. Most

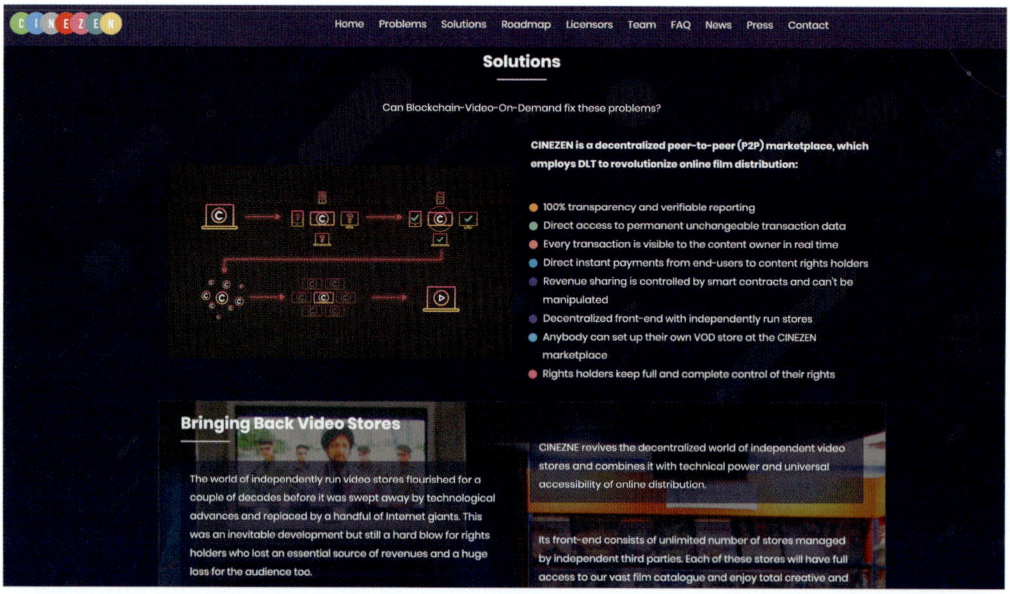

Figure 6.36 CINEZEN is a decentralized platform where everyone can become a video store owner (Cinezen 2020).

Figure 6.37 MINA VR and 360° video production and Cinematic VR (MINA 2018).

Cinematic VR cameras are operated via smartphones, and 360° videos can be viewed via Google Cardboards on smartphones. Smartphone filmmaking as much as 360° filmmaking is driven by exploring the boundaries and new frontiers. It is not of surprise to see smartphone filmmakers like Arnault Labaronne, Max Schleser or Miriam Ross and the creative technologist Camille Baker expand their smartphone filmmaking practice into Mobile XR. Arnault Labaronne is both the artistic director at WondaVR[43] and producer/director at Ideal VR. Miriam Ross's experimental cinematic VR work was featured in the MINA 2018 VR showcase and Camille Baker's research into Mobile XR is outlined in Chapter 4, and she presented at MINA 2011.

In 2017 Mobile Cinematic VR was introduced to the MINA screening programme[44] and in 2018 a dedicated VR showcase was presented at the *Smart Storytelling Day* at Swinburne University of Technology (Melbourne, Australia). Over the last nine years, several AR projects were showcased at MINA,[45] and a number of audio narratives as spatial interventions demonstrate how storytelling can be placed in any location.[46] In this book, Gerdha Cemmar talked about the *Viewfinders* project, which exemplifies how experimental film approaches can be applied to immersive media. The *Viewfinders AR* app enables one to take a filmic experience into the world. Smartphones afford the ability to create new viewing experience in non-cinematic spaces (Figure 6.38).

While there are some limitations to Google Cardboard it is the most accessible form of Cinematic VR (Figure 6.39). As most film festivals focus on Cinematic VR produced on

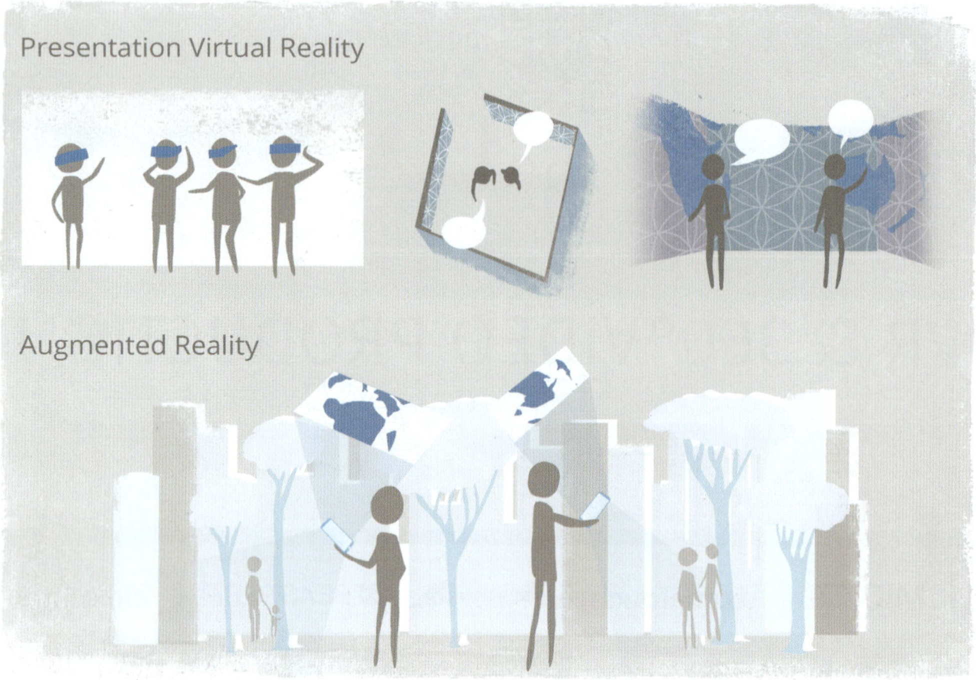

Figure 6.38 Mobile XR (D'Oro 2020).

Figure 6.39 MINA VR showcase Google Cardboard (MINA 2020).

high-end 360° video cameras, the Cinematic VR works selected for MINA were chosen according to its creative capacity and creative strategies for storytelling. Eric Adelheim's *A Little Negro Boy's Prayer* uses a poetic approach and combines landscape VR cinematography with a minimalist animation in the form of a stick figure. *Interwoven: Veganism, Ethics, Economics* by Raul Ramirez uses a smart camera set-up which places the viewer amongst the protagonists into a conversation, which is centred around the topics indicated in the title. Dawen Chan uses a similar approach, but instead of being part of a conversation between two people, the viewer is placed into an argument between the head and heart or in Sigmund Freud's structural model of the psyche's ego, which negotiates between the id, Chan's desires and the super-ego, our institutionalized cultural understanding (Freud 1999). Another example is Miriam Ross's *PND*, an autobiographical exploration of antenatal experiences which is described as 'an alienating period where the pregnant body is measured and constrained in ways that seem at odds with the public emphasis on both the "naturalness" of the event and "pregnancy glam"' (Ross and Neal 2018).[47] The MINA Mobile Cinematic VR category expands smartphone filmmaking's experimentation into the domain of XR.

As an outlook for future research one could explore the smartphone filmmaking modes in the domain of Cinematic VR (CVR) as conceptual parallels between smartphone filmmaking and Mobile Cinematic VR (MCVR) to further develop discussions on community-engaged work, capturing locations or working in the domain of personal or first-person filmmaking. The accessibility and mobility of MCVR demonstrate that individual stories and moving-image arts expand the boundaries of filmmaking in theory and practice through the affordances of mobile media, smartphones and pocket cameras (Figure 6.40).

Figure 6.40 Dr Max Schleser presented a keynote at the International Mobile Storytelling Congress in 2020 at the University of Nottingham, Ningbo, China (IMSC 2020).

With developments towards Creative Arts 4.0[48], Smart Cities and the Internet of things, smartphone filmmaking takes a central position, as a median that can provide novel connections for stories to be created within communities, constructing novel narratives of places and people and continue to drive experimental screen productions. When exploring the technical limitations in the spirit of the mobile filmmaking pioneers, the early Keitai aesthetics can provide a framework for analysis and inspiration. Smartphone filmmaking is driven by accessibility, composed of filmmakers, artists, designer, curators and entrepreneurs, and experimentation. It is this community that is defining creative innovation for a twenty-first-century approach to storytelling and crafting of novel experiences.

Notes

1. Moondog Labs, https://moondoglabs.com/.
2. FiLMiC Pro app, https://www.filmicpro.com/.
3. *Unsane* on IMDb Internet Movie Database, https://www.imdb.com/title/tt7153766/.
4. Festival de Cannes 2018 online, https://www.festival-cannes.com/en/films/les-plus-belles-annees-dune-vie.
5. Berlinale 2019 online, https://www.berlinale.de/en/programm/berlinale_programm/datenblatt.html?film_id=201916356, https://www.berlinale.de/en/archiv/jahresarcLhive/2019/02_programm_2019/02_Filmdatenblatt_2019_201916356.html?openedFromSearch=true#tab=filmStills and https://www.berlinale.de/en/programm/berlinale_programm/datenblatt.html?film_id=201913468.
6. Behind-the-scenes *Experiments I*, https://youtu.be/xL8piHkl3X8.
7. *Experiments II: Full Circle*, https://youtu.be/mCLfFEv_XNw.
8. *Tales from Yarriambiack Shire*, http://bit.ly/YarriambiackTales.
9. Epic Tutorials, https://epictutorials.com/ / https://www.youtube.com/user/EpicTutorialsDotCom.
10. Rob Montgomery, https://www.youtube.com/user/montgomerymultimedia.
11. Smart Film School, https://www.smartfilmschool.com/.
12. Mobile Film School, https://internationalmobilefilmfestival.com/about/mobile-film-school.
13. MoMo Blog, https://momofilmfest.com/newstories/.
14. Cassius Rayner's Budget Gear overview, https://www.mobilefilmmaking.com/budget-gear.
15. Richard Lackey's website, https://www.richardlackey.com/.
16. Mobile Creator Summit, https://www.mobilecreatorsummit.com/.
17. Mobile Creator Podcast, https://www.mobilecreatorpodcast.com/.
18. MINA smartphone filmmaking workshops, https://mina.pro/workshops/.
19. *Spirit of Rangatahi Mobile Filmmaking Workshop*, https://youtu.be/gR7K3z4nURw.
20. *Amazing Day* Global Film Project, https://youtu.be/GiNUWVNtpv8.
21. Behind the Scenes: Capturing Eminem's 'Venom' with Pixel 3, https://www.youtube.com/watch?v=d5aXrvW23WM.
22. John Legend – A Good Night – Behind the Scenes, https://www.youtube.com/watch?v=Wu2tITead8c.

23. Edward Sharpe and the Magnetic Zeros – 'No Love Like Yours' – Behind the Scenes, https://www.youtube.com/watch?v=4Zcss5UiMBY.
24. *High Fantasy*, https://www.imdb.com/title/tt7284066/.
25. MoMo Interview with Kipp Tribble, https://momofilmfest.com/making-char-man-feature-film-shot-smartphone/.
26. *Sickhouse*, https://vimeo.com/ondemand/sickhouse.
27. *#FollowMe* on Instagram, https://www.instagram.com/tv/BxZd8sVgnFr/.
28. *#FollowMe* on YouTube, https://www.youtube.com/watch?v=-zjkkWIyNZk.
29. *This Is Not a Film*, https://www.imdb.com/title/tt1667905/.
30. *Silvered Water, Syria Self-Portrait*, https://www.idfa.nl/en/film/c752cb40-ddaa-4fa6-accd-fefd32b8ac26/silvered-water-syria-self-portrait.
31. *#HashTagOurStories*, https://hashtagourstories.com/.
32. VidCon, https://vidconaustralia.com/agenda/14652/meet-the-stars-of-tiktok/.
33. FiLMiC Fest (https://www.filmicpro.com/filmicfest/mobile2019/) made-on-mobile category was won by Jon Gill with *A Perfectly Healthy Situation*.
34. More information about the album, inclusive of preview links and the statement from the producer, is available on Clan Analogue's web site: https://www.clananalogue.org/featured/ca053-mobile-strategies-various-artists/.
35. https://desistfilm.com/verena-paravel-and-lucien-castaing-taylor-refocusing-the-subject/.
36. LUX https://lux.org.uk/work/ah-humanity and https://www.youtube.com/watch?v=Rc46pQ27uVc.
37. Drone Film Festival http://www.drone-filmfestival.com/.
38. NYC Drone Film Festival https://www.nycdronefilmfestival.com/.
39. Such as Robot Film Festival http://robotfilmfestival.com/, https://spectrum.ieee.org/automaton/robotics/diy/robot-film-festival-watch-all-the-films or ROS Film Festival (http://shortfilms.rosfilmfestival.com), which screened *Robot with A Movie Camera* http://shortfilms.rosfilmfestival.com/video/ew9rnnz3nt59.
40. Fagotto on Cinezen https://beta.cinezen.io/movies/235599?languageId=en.
41. Cinezen https://about.cinezen.io/#solutions.
42. At the Festival de Cannes Film Market, the Marché du Film, the market place for immersive technologies and entertainment launched under the title Cannes XR. Extended reality (XR) is an umbrella term synthesizing augmented, virtual, and mixed reality technologies http://www.marchedufilm.com/en/cannesxr/home.
43. WondaVR is an software to create interactive 360° videos and Cinematic VR.
44. *When We Land* by Ronen Tanchum and *Sonar* by Philipp Maas.
45. Such as *Grey Matter AR* by Karen Vanderborght (2019), Maggie Buxton *Augmenting the Spirit of Place* (2014) or Rewa Wright *Beyond the network and the node: Emergent art practices and the smartphone* (2013)
46. Trudy Lane presented in 2013 *A re-mapping of place and time. A Walk Through Deep Time is a series of walk events and a social practice, which combine to develop an evolving location-aware audio installation* and Hamish Sewell the *Soundtrails project*.
47. MINA Cinematic VR showcase http://mina.pro/screenings/2018-2/.
48. As introduced by Professor Kim Vincs at the Smart Storytelling Day at Swinburne University of Technology, https://youtu.be/LqLpLJ1Jvvw

References

Anderson, Chris. 2004. 'The Long Tail', *Wired Magazine*. Available online: https://www.wired.com/2004/10/tail/.

Baker, Camille, Schleser, Max, Molga, Kasia. 2009. 'Aesthetics of Mobile Media Art', *Journal of Media Practice*, 10 (2&3): 101–22. https://doi.org/10.1386/jmpr.10.2-3.101_1.

Cammaer, Gerda, and Schleser, Max. 2018. 'Viewfinders: A Collaborative Travel Film Project: Seeing the World Through the Lens of the Pocket Camera', in Hannah Brasier, Nicholas Hansen, Kim Munro and Franziska Weidle (eds), *Docuverse: Approaches to Expanding Documentary*, 63–75. Zenodo.

Chicchio. 2019. Personal Communication.

Davis, Hilary, Waycott, Jenny and Schleser, Max. 2019. 'Digital Storytelling: Designing, Developing and Delivering with Diverse Communities', in Satu Miettinen and Melanie Sarantou (eds), *Managing Complexity and Creating Innovation Through Design*, 131–40. London: Routledge Taylor and Francis

De Michiel, Helen and Zimmermann, Patricia. 2017. *Open Space New Media Documentary: A Toolkit for Theory and Practice*. New York: Routledge.

Falt, Chris. 2019. 'Cannes Cinematography Survey: Here's the Cameras and Lenses Used to Shoot 54 Films', *IndieWire*. Available online: https://www.indiewire.com/2019/05/cannes-2019-cameras-lenses-cinematography-survey-1202139519/2/?fbclid=IwAR2aRaK5B4Zgys3Z6ukOgbEjbng-6V7INsRAK3i-dTev4UQDsKmJkYuRNZI.

Fosheim, Ricky. 2014. 'Shooting a Feature on an iPhone', in *Production Slate – American Cinematographer*, 20–6. August 2014. Available online: https://www.scribd.com/document/274535068/American-Cinematographer-August-2014.

Freud, Sigmund. 2019/1923. *The Ego and the ID*. New York City: Simon and Schuster.

Handke, Sebastian. 2009. 'Iran ist wie ein Vulkan', *Zeit*. Available online: https://www.zeit.de/online/2009/33/iran-handy-film.

Klebanov, Sam. 2020. *Cinezen Blockchained Entertainment AB*. https://about.cinezen.io/.

Leu, Allison. 2016. 'Grimes Released 7 New Videos Shot on iPhone', *Seenit*. Available online: https://seenit.io/blog/2016/10/07/grimes-released-7-videos/.

McClintock, Pamela. 2018. 'Box-Office Preview: "Pacific Rim Uprising" Set to Stomp Past "Black Panther"', *The Hollywood Reporter*. Available online: https://www.hollywoodreporter.com/news/box-office-preview-pacific-rim-uprising-set-stomp-past-black-panther-1096491.

Paquet, Darcy. 2011. 'Night Fishing', *Screen Daily*. Available online: https://www.screendaily.com/night-fishing/5023095.article.

Phelan, David. 2019. 'Movie Shot on iPhone by Oscar-Winning Director Premieres At Cannes', *Forbes Magazine*. Available online: https://www.forbes.com/sites/davidphelan/2019/05/25/movie-shot-on-iphone-from-oscar-winning-director-premieres-at-cannes-film-festival-filmic-pro/#bf732211b4ed.

Ross, Miriam and Glen, Maddy. 2014. 'Vertical Cinema: New Digital Possibilities', *Rhizomes*, [S. l.], 26. Available online: http://www.rhizomes.net/issue26/ross_glen.html (Accessed 11 November 2017).

Ross, Miriam and Neal, Dave. 2018. Mobile Framing: Vertical Videos from User-Generated Content to Corporate Marketing, in Max Schleser and Marsha Berry (eds), *Mobile Story Making in an Age of Smartphones*, 1–7. London: Springer.

Schleser, Max. 2011. *Mobile-Mentary: Mobile Documentaries in the Mediascape*. Saarbrücken: LAP Lambert Academic Publishing.

Schleser, Max. 2012. 'Collaborative Mobile Phone Film Making', in E.-J. Milne, Claudia Mitchell and Naydene de Lange (eds), *Handbook of Participatory Video*, 397–411. Lanham: AltaMira Press.

Schleser, Max. 2014. 'A Decade of Mobile Moving-Image Practice', in Gerhard Goggin and Larissa Hjorth (eds), *The Routledge Companion to Mobile Media*, 157–71. Routledge: New York.

Schleser, Max. 2018. '24Frames 24Hours: An Emerging Documentary Form: Workshop-Generated Videos', in Adrian Miles (ed.), *Digital Media and Documentary: Antipodean Approaches*, 101–10. London: Palgrave.

Schleser, Max. 2018. 'Smart (Phone) Filmmakers >> Smart (Political) Actions', in Marcus Bohr and Basia Sliwinska (eds), *The Evolution of the Image: Political Action and the Digital Self*, 114–26. London: Routledge.

Snowdon, Peter. 2014. 'The Revolution Will be Uploaded: Vernacular Video and the Arab Spring', *Culture Unbound – Journal of Current Cultural Research*, 6: 401–29.

Snyder, Gabriel. 2007. 'What's in a Name? Why Filmmakers Use Pseudonyms'. *Slate*. Available online: https://slate.com/news-and-politics/2007/01/why-filmmakers-like-steven-soderbergh-use-pseudonyms.html.

Spangler, Todd. 2019. 'Tye Sheridan Starring in Survival Thriller 'Wireless' From Steven Soderbergh for Quibi',. *Variety* (14 November 2019). https://variety.com/2019/digital/news/quibi-wireless-tye-sheridan-steven-soderbergh-1203405209/.

Sterkenburg, Tessa. 2012. 'What Is the Future of Mobile ?', *The Next Web* Issue v0.9. Available online: https://thenextweb.com/magazine/2012/09/25/issue-v0-9-2/.

Tang, Estelle. 2016. 'This Is Why Grimes Made Her Own Music Videos Using Only Her iPhone – It's all About Creative Freedom – And Not Having To Work With Sexist Dudes', *Elle*. Available online: https://www.elle.com/culture/music/interviews/a40368/grimes-acid-reign-chronicles/.

Watercutter, Angela. 2015. 'Tangerine Is Amazing, But Not Because of How They Shot It', *Wired Magazine*. Available online: https://www.wired.com/2015/07/tangerine-iphone/.

Zoller, Matt and Carvajal, Nelson. 2013. 'Video Essay: Peter Andrews: The Soderbergh Vision', *IndieWire*. Available online: https://www.indiewire.com/2013/03/video-essay-peter-andrews-the-soderbergh-vision-133830/.

Films

#FollowMe. 2019. Dir. Nicolaas Veul. Netherlands: VPRO.

#MyEscape. 2016. Dir. Elke Sasse. Germany: Berlin Producers and Deutsche Welle.

#Nucleus. 2019. Dir. Max Schleser. New Zealand: Goethe Institut.

#randomaccessmemory. 2017. Dir. William Brown. UK: begstealborrow films.

#ScratchTheFate. in production. Dir. Luis Eduardo Juarez. United States: independent production.

7/11. 2014. Dir. Beyoncé Knowles and Todd Tourson. United States: EMI Music Publishing.

9 Rides. 2016. Dir. Michael A. Cherry. United States: Cherry Entertainment, Datari Turner Productions, JCaldwell Productions.

24 Frames 24 Hours. 2014–2017. Dir. Max Schleser and international participants. international: Massey University.
A Good Night. 2018. Dir. Mishka Kornai. United States: Whitelist.tv.
A Little Negro Boy's Prayer. 2018. Dir. Eric Adelheim. Guadalupe: defocus films.
A Perfectly Healthy Situation. 2019. Dir. Jon Gill. UK: Playful Comm.
Ah Humanity! 2015. Dir. Verena Paravel and Lucien Castaing-Taylor. United States: LUX.
Amazing Day. 2016. Dir. Coldplay. international: Parlophone Records and The Firepit.
B.A.N. (Official Vertical Video). 2018. Dir. Stephen Garnett. United States: WMG.
Batman v Superman: Dawn of Justice. 2017. Dir. Zack Snyder. United States: Warner Brothers and Atlas Entertainment.
Blair Witch Project. 1999. Dir. Daniel Myrick and Eduardo Sánchez. United States: Haxan Films.
Blue Moon. 2018. Dir. Stefen Harris. New Zealand: Dark Horse Films.
Char Man. 2019. Dir. Kurt Ela and Kipp Tribble. United States: MRP Entertainment, GK & K Productions.
Détour. 2017. Dir. Michel Gondry. France: Apple.
Ephemera. 2018. Dir. Angelo Chiacchio. Italy: independent production.
Eternal Sunshine of the Spotless Mind. 2004. Dir. Michel Gondry. France: Focus Features, Anonymous Content, This Is That Productions.
Eva.Stories. 2019. Dir. Mati Kochavi and Maya Kochavi. Ukraine: Colourfilm, K's Gallery and POV Production.
Experiments. 2019. Dir. Donghoon Jun and James Thornton. United States: Incite and Apple.
Fagotto. 2018. Dir. Boris Gouts. Russia: Studio Moloch.
Fly Away. 2014. Dir. Luis Eduardo Juarez. United States: Christafari.
High Fantasy. 2017. Dir. Jenna Cato Bass. South Africa: Big World Cinema, Deal Productions and Fox Fire Films.
High Flying Bird. 2019. Dir. Steven Soderbergh. United States: Harper Road Films, Extension 765 and Netflix.
Hinterlands. 2016. Dir. Scott Barley. UK: independent production.
Homemade Spacecraft. 2012. Dir. Luke Geissbühler. United States: Brooklyn Space Project.
Hungerlohn am Fließband: Wie Tarife ausgehebelt werden (= Starvation Wages: How Pay Scales Are Cancelled. 2013. Dir. Jürgen Rose. Germany: SWR – Südwest Rundfunk.
I Play with the Phrase Each Other. 2014. Dir. Jay Alvarez. United States: Studio Mist.
In Response We Closed Flinders. 2015. Dir. Alex Dick. Australia: independent production.
Interwoven: Veganism, Ethics, Economics. 2018. Dir. Raul Ramirez. Canada: VR Vegan.
Letters to Ariadne. 2016. Dir. William Brown. UK: begstealborrow films.
Life in a Day. 2011. Dir. Kevin Macdonald. international: YouTube and National Geographic Films.
Midnight Traveler. 2019. Dir. Hassan Fazili. United States, United Kingdom, Qatar, Canada: Old Chilly Pictures.
Mobile Strategies: Battery-Powered Sonics. 2019. Produced and curated by Martin K. Koszolko. Australia: Clan Analogue.
Mood Indigo. 2013. Dir. Michel Gondry. France: Brio Films, StudioCanal, Scope Pictures and France 2 Cinéma.
Moravia Off. 2017. Dir. Luca Lancise. Italy: Istitu Luce, Stemal Entertainment.
Night Fishing (Paranmanjang). 2011. Dir. Park Chan-wook and Park Chan-kyong. South Korea: Moho Films.

No Love Like Yours. 2016. Dir. Olivia Wilde. United States: Anonymous Content.
Ocean's (film series). 2001, 2004, 2007, 2018. Dir. Steven Soderbergh and Gary Ross. United States: Jerry Weintraub Productions, Section Eight Productions and Village Roadshow Pictures.
Oceans (Where Feet May Fail). 2014. Dir. Luis Eduardo Juarez. United States: Christafari.
Olive. 2013. Dir. Hooman Khalili and Pat Gilles. United States: CaveScribe.
Out of Sight. 1998. Dir. Steven Soderbergh. United States: Jersey Films and Universal Pictures.
Pangea Day. 2018. Dir. Jehane Noujaim. international: Nokia.
PND. 2018. Dir. Miriam Ross. New Zealand: independent production.
POCKET. 2019. Dir. Mishka Kornai and Wechter. United States: Pickpocket, The Whitelist Collective.
Puhelinkoppi [1882–2007]. 2010. Dir. Hope Tucker. Finland and United States: The Obituary Project.
Reel Health. 2011. Dir. Joanna Ong. Tanzania: Tigo.
Saudi Runaway. 2020. Dir. Susanne Regina Meures. Switzerland and Germany: National Geographic.
Selfie. 2014. Dir. William Brown. UK: begstealborrow films.
Selfie. 2018. Dir. Agostino Ferrente. Italy: Magnéto Prod, ARTE, Casa delle Visioni.
Sex, Lies, and Videotapes. 1989. Dir. Steven Soderbergh. United States: Outlaw Productions.
Sickhouse. 2016. Dir. Hannah Macpherson. United States: Indigenous Media.
Silvered Water: Syria Self-Portrait. 2014. Dir. Ossama Mohamed and Wiam Simav Bedirxan. Syria: Les Films d'Ici, Proaction Film.
Sketch-Three: Avant-Garde (*R.P.M. 2*). 2014. Dir. Ryan Fox. United States: independent production.
Sleep Has Her House. 2017. Dir. Scott Barley. UK: Ether Films.
Snow, Steam, Iron. 2017. Dir. Zack Snyder. United States: Vero.
Snowbird. 2016. Dir. Sean Baker. United States: Kenzo.
Some Postman. 2008. Dir. Grant Marshal. United States: Film Headquarters, The Orchard Music and EMI Music Publishing.
Speech Marks. 2004. Dir. Steve Hawley. UK: independent production.
Spirit of Rangatahi Mobile Filmmaking Workshop. 2014. Dir. Max Schleser. New Zealand: MINA and Massey University.
Tales from Yarriambiack Shire. 2019. Dir. Max Schleser and Hilary Davis. Australia, Social Innovation Research Institute.
Tangerine. 2015. Dir. Sean Baker. United States: Magnolia Pictures.
Tehran without Permission (= Tehran Bedoune Mojavez). 2009. Dir. Sepideh Farsi. Iran: Rêves d'Eau Productions and Shariati.
The Best Years of a Life (= Les plus belles années d'une vie). 2019. Dir. Claude Lelouch. France: Les Films 13, Davis-Films, France 2 Cinéma.
The Big City. 2018. Dir. Evan Luchkow. Canada: Aeon.
The Green Ray. 2017. Dir. Scott Barley. UK: independent production.
The Informant! 2009. Dir. Steven Soderbergh. United States: Participant Media, Groundsell Productions, Section Eight.
The Life and Death of the iPhone. 2015. Dir. Paul Trilli. United States: Cameo.
The New Hope. 2015. Dir. William Brown. UK: begstealborrow films.
The Uprising. 2013. Dir. Peter Snowdon. Belgium: Rien à Voir Production and Third Films.

This Is Not a Film. 2011. Dir. Jafar Panahi. Iran: Jafar Panahi Film Productions.
Throwback '89. 2019. Dir. Ricarda Saleh, Nil Idil Çakmak. Germany: NDR – Norddeutscher Rundfunk.
Uneasy Lies the Mind. 2014. Dir. Risky Fosheim. United States: Detention Films and All Mod Cons.
Unsane. 2018. Dir. Steven Soderbergh. United States: Regency Enterprises, Extension 765 and 20th Century Fox.
V2. Escape From Hell. in production. Dir. Timur Bekmambetov. Russia, Bazelevs: Voenfilm and MYS Media.
Venom. 2018. Dir. Rich Lee and James Larese. United States: Aftermath Records.
Vertical Video Syndrome – A PSA. 2013. Dir. The Glove and Boots Stash. United States: The Glove and Boots Stash. https://vimeo.com/313458699.
Viewfinders. 2014. Dir. Gerda Cammaer and Max Schleser. international: independent production. http://www.viewfinders.gallery/.
Vladimir and William. 2018. Dir. William Brown. UK: begstealborrow films.
We Look Good in Death. 2019. Dir. Boris Gouts. Russia: Igor Mishin & Maxim Mussel.
Without Me (Vertical Video). 2018. Dir. Colin Tilley. United States: Capitol Records, LLC.

Index